Hydrology and Hydraulic Systems

Hydrology and Hydraulic Systems

Francis Stone

STATES
ACADEMIC PRESS
www.statesacademicpress.com

Hydrology and Hydraulic Systems
Francis Stone
ISBN: 978-1-63989-279-2 (Hardback)

STATES
ACADEMIC PRESS

Published by States Academic Press,
109 South 5th Street,
Brooklyn, NY 11249, USA

Cataloging-in-Publication Data

Hydrology and hydraulic systems / Francis Stone.
p. cm.
Includes bibliographical references and index.
ISBN 978-1-63989-279-2
1. Hydrology. 2. Hydraulic control. 3. Hydraulic machinery. 4. Hydraulics.
5. Hydrography. 6. Water. I. Stone, Francis.
GB661.2 .H93 2022
551.48--dc23

For more information regarding States Academic Press and its products, please visit the publisher's website www.statesacademicpress.com

Table of Contents

Preface

The branch of science which focuses on the distribution, management and movement of water on Earth and other celestial bodies is known as hydrology. This includes a detailed analysis of water resources, water cycle and environmental watershed sustainability. The discipline of hydrology is divided into various sub-disciplines such as surface hydrology, hydrometeorology, drainage-basin management and hydrogeology. Hydrology finds applications in a wide variety of fields such as disaster management, public health, power generation, civil engineering, etc. Systems and mechanism which make use of fluid power for performing work are called hydraulic systems. They are capable of transferring large amounts of power through very small hoses and tubes. The key components of a hydraulic system are hydraulic pumps, actuators, control valves, accumulators, reservoirs, hydraulic fluids, filters, seals and pipes. The topics included in this book on hydrology and hydraulic systems are of utmost significance and bound to provide incredible insights to readers. It attempts to understand the multiple branches that fall under the discipline of hydrology and hydraulic systems, and how such concepts have practical applications. Those in search of information to further their knowledge will be greatly assisted by this book.

Given below is the chapter wise description of the book:

Chapter 1- The scientific study of water present on the surface of Earth and beneath the surface of Earth is known as hydrology. It deals with the physical and chemical properties of water, the movement, distribution and management of water and the relationship of water with living organisms. This is an introductory chapter which will introduce briefly all the significant aspects of hydrology.

Chapter 2- The cycle that explains the continuous circulation of water from the time it falls to the ground until it returns to the atmosphere is known as the hydrologic cycle. This cycle also involves the energy exchange which leads to temperature changes. This chapter has been carefully written to provide an easy understanding of the varied facets of water cycle.

Chapter 3- An aquifer is the saturated zone beneath the water table. Aquifers are permeable and porous so that the water can travel through. They naturally filter the groundwater by making it pass through small pores. This chapter covers the types and properties of aquifers. The topics elaborated in this chapter will help in gaining a better perspective about the aquifers.

Chapter 4- The water present under the surface of the Earth in the cracks and spaces of soil is known as groundwater. The water located on the surface of Earth in the form of streams, rivers, lakes, reservoirs, wetlands, etc. is known as surface water. The chapter closely examines the key concepts of groundwater and surface water to provide an extensive understanding of the subject.

Chapter 5- The flow of water over the surface of Earth due to rain, snowmelt or other sources is known as surface runoff. It occurs when the water cannot rapidly infiltrate in the soil. This chapter has been carefully written to provide an easy understanding of the varied facets of surface runoff such as first flush, meltwater runoff, rainfall runoff, stormwater runoff, urban runoff, etc.

Chapter 6- Statistical evaluation in hydrology includes the statistical procedures for evaluating the daily and monthly predictions of hydrologic models. This chapter elaborates statistical analysis, probability distribution and statistical hypothesis testing. All the diverse principles of statistical evaluation have been carefully analyzed in this chapter.

At the end, I would like to thank all those who dedicated their time and efforts for the successful completion of this book. I also wish to convey my gratitude towards my friends and family who supported me at every step.

Francis Stone

1

Introduction to Hydrology

The scientific study of water present on the surface of Earth and beneath the surface of Earth is known as hydrology. It deals with the physical and chemical properties of water, the movement, distribution and management of water and the relationship of water with living organisms. This is an introductory chapter which will introduce briefly all the significant aspects of hydrology.

Hydrology is study of water in the broadest sense. It encompasses the occurrence, distribution and circulation of water, its physical and chemical properties and its relation to living things. Water is a chemical compound with the chemical formula H_2O. A water molecule contains one oxygen and two hydrogen atoms that are connected by covalent bonds. Water is a liquid at standard ambient temperature and pressure, but it often co-exists on earth with its solid state, ice, and gaseous state, steam (water vapour).Water covers 71% of the earth's surface, and is vital for all known forms of life. On earth, 96.5% of the planet's water is found in seas and oceans, 1.7% in groundwater, 1.7% in glaciers and the ice caps of Antarctica and Greenland, a small fraction in other large water bodies, and 0.001% in the air as vapour, clouds (formed of solid and liquid water particles suspended in air), and precipitation. Only 2.5% of the earth's water is fresh water, and 98.8% of that water is in ice and groundwater. Less than 0.3% of all freshwater is in rivers, lakes, and the atmosphere, and an even smaller amount of the earth's freshwater (0.003%) is contained within biological bodies and manufactured products. Water on earth moves continually through the water cycle of evaporation and transpiration (evapotranspiration), condensation, precipitation, and runoff, usually reaching the sea. Evaporation and transpiration contribute to the precipitation over land.

Importance of Water for Life and Environment

All the water in its various forms, present in all the components of earth's environment together constitutes the hydrosphere. Most of the water is present as liquid water on the earth's surface and some liquid water is present underground. Apart of total water is present as snow or ice on the earth's surface while a substantial part of water in earth's environment is also present as water vapor in the atmosphere. General features of various parts of the earth's hydrosphere are as given below:

- Oceans: Major part of water is present in the oceans of the earth. Average depth of oceans is about 3.7 kilometres and about 1300 million cubic kilometres water is present in oceans.
- Ice sheets: Substantial quantity of water, about 24 million cubic kilometres, is present as solid in the ice sheets of Earth. About 90% of the volume of such water is found in Antarctica.
- Groundwater: About 24 million cubic kilometre water is present under the ground surface at depths of up to two kilometres.
- Lakes and rivers: On the land surface, approximately 0.18 million cubic kilometer water is present in lakes while about 0.002 million cubic kilometer water is found in rivers.

- Atmospheric moisture: The amount of water present as water vapor in the atmosphere is about 0.013 million cubic kilometer.
- Biological water: In addition to above categories, about 0.001 million cubic kilometer water is contained in the bodies of living organisms.

The earth did not have any hydrosphere in the beginning. It is thought that hydrosphere emerged as the result of processes taking place in the lithosphere. These processes released a substantial quantity of water vapor and juvenile waters during the geological history of earth. Further, the amounts of water present in oceans, as ice and in atmosphere have fluctuated with appearance and disappearance of major periods of glaciations on the earth. Palaeogeographic data indicates that the level of oceans on earth had declined by more than 100 meters during the age of greatest glaciations during Quaternary period. It is estimated that if ice sheets present on earth today were to melt completely, the level of oceans of earth would rise by about 66 meters. Most important feature of global environment of earth is the hydrological cycle which determines the distribution of water on earth's surface and in the atmosphere. The water is evaporated from surface of water bodies like oceans, lakes, rivers etc. as water vapor into atmosphere. It is also absorbed by plants from soil and lost as vapor to atmosphere through transpiration. The water vapor condenses in the atmosphere to form precipitation and thus water is returned from the atmosphere to the surface of Earth. Cyclic movement of water in different components of the global environment is termed hydrological cycle.

The water cycle, or hydrologic cycle, is a continuous process by which water is purified by evaporation and transported from the earth's surface (including the oceans) to the atmosphere and back to the land and oceans. All of the physical, chemical and biological processes involving water as it travels its various paths in the atmosphere, over and beneath the earth's surface and through growing plants, are of interest to those who study the hydrologic cycle.

There are many pathways the water may take in its continuous cycle of falling as rainfall or snowfall and returning to the atmosphere. It may be captured for millions of years in polar ice caps. It may flow to rivers and finally to the sea. It may soak into the soil to be evaporated directly from the soil surface as it dries or be transpired by growing plants. It may percolate through the soil to ground water reservoirs (aquifers) to be stored or it may flow to wells or springs or back to streams by seepage. The cycle for water may be short, or it may take millions of years.

People tap the water cycle for their own uses. Water is diverted temporarily from one part of the cycle by pumping it from the ground or drawing it from a river or lake. It is used for a variety of activities such as households, businesses and industries; for irrigation of farms and parklands; and for production of electric power. After use, water is returned to another part of the cycle: perhaps discharged downstream or allowed to soak into the ground. Used water normally is lower in quality, even after treatment, which often poses a problem for downstream users.

The hydrologist studies the fundamental transport processes to be able to describe the quantity and quality of water as it moves through the cycle (evaporation, precipitation, stream flow, infiltration, groundwater flow, and other components). The engineering hydrologist, or water resources engineer, is involved in the planning, analysis, design, construction and operation of projects for the control, utilization, and management of water resources. Water resources problems are also the concern of meteorologists, oceanographers, geologists, chemists, physicists, biologists, economists, and political scientists, specialists in applied mathematics and computer science, and engineers in several fields.

Branches of Hydrology

Surface Hydrology

It is the study of bodies of water on or near the earth's surface. Rivers, dams, lakes, and reservoirs are all part of this area of study, which further includes the systems used in recreational activity and transportation. Surface hydrology addresses issues pertaining to eroding soils and streams due to surface flow. Flooding, nutrient runoff, and pollutants are a few of the effects addressed, as well as the destruction of civil constructions such as dams. Methods of hydraulic and hydrologic design regulation are also undertaken in this field of study, as researchers simulate the long and short-term effects of anthropogenically manipulated surface water forms.

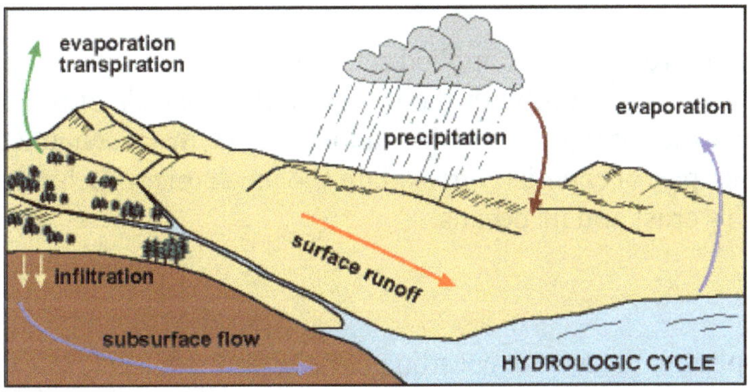

Hydromorphology

Hydromorphology is the study of the physical characteristics of bodies of water on the Earth's surface, including river basins, channels, streams, and lakes. Water quality, levels of pollution, and biological components needed for ecological system maintenance are a few areas assessed when classifying water systems. Hydromorphology studies the dynamics of groundwater flow into channels, lakes, and streams. It measures flow patterns and geometry as well as routing flows to avoid flooding or drought.

Hydrometeorology

Hydrometeorology is the study of the transfer of water and energy between land and water body surfaces and the lower atmosphere. Hydrometeorology incorporates meteorology to solve hydrological problems. These problems include forecasting flood or drought, or determining water resources and the safety of dams. Hydrometeorologists try to determine, through empirical data or theory, how the dynamics of water in the atmosphere affect the greatest levels of precipitation reaching the ground. The domain of hydrometeorology in the physical sciences is not very clearly defined, as it involves cloud physics, climatology, weather forecasting, and hydrology, to name a few.

Hydrogeology is the study of the distribution and movement of water in aquifers and shallow porous media—that is, the porous layers of rock, sand, silt, and gravel below the Earth's surface. Hydrogeology examines the rate of diffusion of water through these media as the water moves down its energy gradient. The flow of water in the shallow subsurface is also pertinent to the fields of soil science, agriculture, and civil engineering. The flow of water and other fluids (hydrocarbons and geothermal fluids) in deeper formations is relevant to the fields of geology, geophysics, and petroleum geology. Geohydrology is the area of geology that deals with the distribution and movement of groundwater in the soil and rocks of the Earth's crust. The term geohydrology is often used interchangeably.

Ecohydrology is the study of ecological processes in the hydrologic cycle. As these processes occur in the soil and plant foliage, ecohydrologists study how the hydrologic system affects plant physiology, soil moisture, and plant diversity and spatial orientation in various regions over a period of time. Ecohydrology has four main components: infiltration of precipitation into the soil, evapotranspiration, leakage of water into deeper portions of the soil not accessible to the plant, and runoff from the ground surface.

Forest Hydrology

Forest hydrology is an interdisciplinary applied science that intersects the disciplines of hydrology and forestry. It is primarily involved with the research and analysis of the effects of forest cover and changes of land use on water yield, quality and timing. Hydrogeochemistry is the science that uses the tools and principles of chemistry to explain the mechanisms behind major geological systems such as the earth's crust and its oceans.

Hydrography

Hydrography as it applies to the area of coastlines can be defined as: (1) The description and study of seas, lakes, rivers and other waters. (2) The science of locating aids and dangers to navigation. (3) The description of physical properties of the waters of a region'. The term 'Hydrography' as it applies to the area of reclamation can be defined as ' Scientific study of physical aspects of all waters on the Earth's surface. Water features in 7.5-minute quads include lakes, shorelines, and drainage routing'.

Karst Topography

Karst is a geological formation shaped by the dissolution of a layer or layers of soluble bedrock usually carbonate rock such as limestone or dolomite, but also in gypsum. It has also been documented for weathering-resistant rocks, such as quartzite, given the right conditions. Karst topography is a geological formation shaped by the dissolution of a layer or layers of soluble bedrock usually carbonate rock such as limestone or dolomite, but also in gypsum. It has also been documented for weathering-resistant rocks, such as quartzite, given the right conditions. Subterranean drainage may limit surface water with few to no rivers or lakes.

Many karst regions display distinctive surface features, with cenotes and sinkholes (also called dolines) being the most common. However, distinctive karst surface features may be completely absent where the soluble rock is mantled, such as by glacial debris or confined by one or more superimposed non-soluble rock strata. Some karst regions include thousands of caves, although evidence of caves large enough for human exploration is not a required characteristic of karst.

Hydrogeophysics

This branch has emerged over the last decade as one of the more challenging disciplines in near surface geophysics, aiming to improve the simultaneous use of geophysical and hydrogeological measurements. It can be described as the use of geophysical measurements for mapping subsurface features, estimating properties and monitoring processes that are important to hydrological studies, such as those associated with water resources, seepage throughout the vadose zone, contaminant transport, and ecological/climate investigations.

Vadose Zone Hydrology

It is also termed the unsaturated zone is the part of Earth between the land surface and the top of the phreatic zone i.e. the position at which the groundwater (the water in the soil's pores) is at atmospheric pressure ("vadose" is from the Latin for "shallow"). Hence the vadose zone extends from the top of the ground surface to the water table. Water in the vadose zone has a pressure head less than atmospheric pressure, and is retained by a combination of adhesion (funiculary groundwater), and capillary action (capillary groundwater). If the vadose zone envelops soil, the water contained therein is termed soil moisture.

In fine grained soils, capillary action can cause the pores of the soil to be fully saturated above the water table at a pressure less than atmospheric. In such soils, therefore, the unsaturated zone is the upper section of the vadose zone and not identical to it. Movement of water within the vadose zone is studied within soil physics and hydrology, particularly hydrogeology, and is of importance to agriculture, contaminant transport, and flood control. The Richards equation is often used to mathematically describe the flow of water, which is based partially on Darcy's law. Groundwater recharge, which is an important process that refills aquifers, generally occurs through the vadose zone from precipitation.

Coastal Zone Hydrology

There is no standard definition of what constitutes a 'coastal zone', but it could be said to be the interface between the land and water. These zones are dynamic areas with frequently changing biological, chemical and geological attributes and are of high economic significance, which is often subject to fast economic development, large population migrations and urban development.

Island Zone Hydrology

Island zone hydrology is a lens is a convex layer of fresh groundwater that floats on top of denser saltwater. It arises when rainwater seeps down through a soil surface and then gathers over a layer of seawater at or down to about five feet below sea level. Freshwater lenses are often found on small coral or limestone islands and atolls, where wells dug into them may be the only natural source of potable water.

Arid Zone Hydrology

Arid areas have distinctive hydrological features substantially different from those of humid areas. The term arid refers to severe lack of water in a region. The shortage of water prevents or hinders development and growth of animal and plant life. Arid areas lack vegetation and they are also called deserts. Arid zones are usually areas of scarce hydrological data. Any study requires an iterative approach to be adopted to develop water resources, with initial information often derived from other geographical regions which have similar aridity characteristics. Arid zones present special problems to hydrologists. Describing the hydrological balance is difficult as very little of the potential evaporation demand is satisfied. Only a small proportion of the rainfall becomes runoff except in extreme storms. The amount of rainfall tends to be highly variable in arid regions both in time and space. This leads to very high variability of runoff. Water, especially groundwater, is the key to life in semi-arid and arid zones.

Wadi Hydrology

Wadi is the Arabic word for ephemeral water courses in the arid regions, and they are a vital source of water in most arid and semi-arid countries. Catastrophic flash floods occurring in wadis are, on one hand, a threat to many communities and, on the other, major groundwater recharge sources after storms.

In regions of the Middle East and North Africa, a stream bed or channel that only carries water during the rainy season. In the southwest United States, the equivalent terms would be arroyo or wash. Wadi Hydrology has emerged as a distinct scientific area within the last decade, due mainly to the initiative of a small number of individuals within and without the Arab region, and the active support of UNESCO, assisted by ACSAD and ALECSO. This has been due to the recognition that the hydrology of arid and semi-arid areas is very different from that of humid areas and raises important scientific, technical and logistical challenges, and that an improved science base is essential to meet current and future needs of water management. By definition, water is a scarce resource in arid regions, and most countries of the Arab region and other arid and semi-arid areas are facing severe pressures due to limited water resources. Wadi bed infiltration has an important effect on flood propagation, but also provides recharge to alluvial aquifers. The balance between distributed infiltration from rainfall and wadi bed infiltration is obviously dependant on local conditions, but soil moisture observations.

Spate Hydrology

Spate hydrology is characterized by a great variation in the size and frequency of floods, which directly influence the availability of water for agriculture in any one season. Spate floods can have very high peak discharges and are usually generated in wadi catchments by localized storm rainfall. Crop production varies considerably because of the large variation in wadi runoff from year to year, season to season and day to day. Hydrological and sediment transport data are needed to design improved water diversion structures and canals in spate schemes.

Engineering Hydrology

It uses hydrologic principles in the solution of engineering problems arising from human exploitation of water resources of the earth. The engineering hydrologist, or water resources engineer, is involved in the planning, analysis, design, construction and operation of projects for the control, utilization and management of water resources. Hydrologic calculations are estimates because mostly the empirical and approximate methods are used to describe various hydrological processes. This branch helps in the following ways:

- Hydrology is used to find out maximum probable flood at proposed sites e.g. dams.
- The variation of water production from catchments can be calculated and described by hydrology.
- Engineering hydrology enables us to find out the relationship between a catchment's surface water and groundwater resources.
- The expected flood flows over a spillway, at a highway Culvert, or in an urban storm drainage system can be known by this very subject.
- It helps us to know the required reservoir capacity to assure adequate water for irrigation or municipal water supply in droughts condition.
- It tells us what hydrologic hardware (e.g. rain gauges, stream gauges etc.) and software (computer models) are needed for real-time flood forecasting.

Hydraulics

The branch on Hydraulics deals with the study of fluids: their behavior, motion and interaction of fluids with other bodies. Technically fluids include liquids and gases, but from the perspective of Hydraulics in Civil Engineering the term fluid generally means a liquid and water in particular. The topics covered include channel versus pipe flow, continuity equation, pressure distribution, specific energy, specific force, trapezoidal and circular channels, hydraulic exponent, measuring flumes, critical depth flumes, weir introduction, control structures, proportional weirs, flow over weir's, types of broad crested weirs, bear trap, sluice gate, brink depth, outlets, modules, canal lining, hydraulic jumps, spillways, surges, dam break analysis etc.

Hydraulic Engineering

Hydraulic Engineering covers the planning, management, design and operation of water supply and distribution systems, flood control and flood hazard mapping, hydrologic and hydraulic aspects of environmental issues, and application of remotely-sensed data to hydrologic and environmental problems. Hydraulic engineering is the application of fluid mechanics to the liquid earth.

Stochastic Hydrology

The objective of this course is to introduce the concepts of probability theory and stochastic processes with applications in hydrologic analysis and design. Modeling of hydrologic time series with specific techniques for data generation and hydrologic forecasting are dealt with and case study applications discussed.

Urban Hydrology

Urban hydrology is a science, part of land hydrology investigating the hydrological cycle, water regime and quality in urbanized territory. Urban hydrology is an applied science that will have an increasing role to play in the sustainability of human societies. Facing present growth of urban population, it is increasingly difficult to find and utilize new sources of water necessary to satisfy growing water demand.

Flood Hydrology

Flood hydrology entails the calculation of flood peaks or flood hydrographs for observed floods or for design floods for specified return periods. This branch of hydrology involves both the separate actions and the interactions of forces and parameters in geomorphology, hydraulics, hydrology, physical geography, economics, engineering and planning. The flood hydrology deals with two principal areas:

- Determination of the upper limit or probable maximum flood potential at a dam site so that dams whose failure would result in loss of human life or widespread property damage can be designed to safely accommodate this flood without failure.
- Determination of more commonly occurring floods for use in the design of diversion dams; very low hazard storage dams; cross drainage facilities for the extensive canals, aqueducts, and roads associated with the delivery of water to users; and for diversion of flood waters that may occur during the construction of dams. Flood hydrology studies support the planning, design, construction, and operation of flood control projects.

There are three major technical aspects in flood hydrology. They are: (1) Hydrometeorology related to probable maximum precipitation determinations, (2) Probable maximum flood hydrograph determinations, and (3) Statistics and probabilities relating to the magnitude and frequency of flood flows.

Hydrometry

Field hydrometry involves the collection of all types of hydrological data involving water level, flow, rainfall and many other variables. The following aspects are covered under this branch:

- Continuous Flow Measurement.
- Spot Flow Measurement.
- Water Level.
- Rainfall and Climate.
- Water Quality.

Highway Hydrology

Hydrologic methods of primary interest are frequency analysis for analyzing rainfall and ungaged data; empirical methods for peak discharge estimation; and hydrograph analysis and synthesis. The estimation of peak discharges of various recurrence intervals is one of the most common problems faced by engineers when designing for highway drainage structures. For the highway designer, the primary focus of hydrology is the water that moves on the earth's surface and in particular that part that ultimately crosses transportation arterials (i.e., highway stream crossings). A secondary interest is to provide interior drainage for roadways, median areas, and interchanges. By application of the principles and methods of modern hydrology, it is possible to obtain solutions that are functionally acceptable and form the basis for the design of highway drainage structures.

Wetland Hydrology

Wetland hydrology is the study of the movement of water in and out of the wetland ecosystem. In wetlands the presence of water is the overwhelming characteristic of the ecosystem. Wetlands are a unique hydrologic feature of the landscape. One particularly important attribute is their position as the transition zone between aquatic and terrestrial ecosystems. Wetlands share aspects of both aquatic and terrestrial environments because of this position. On one hand, most freshwater and marine aquatic environments, such as lakes, rivers, estuaries, and oceans, are characterized as having permanent water. On the other hand, terrestrial environments are generally characterized as having drier conditions, with an unsaturated (vadose) zone present for most of the annual cycle. Wetlands thus occupy the transition zone between predominantly wet and dry environments. A diagnostic feature of wetlands is the proximity of the water surface (or water table below the surface) relative to the ground surface. In freshwater and marine aquatic habitats, the water surface lies well above the land surface, while in terrestrial environments it lies some distance below the root zone as a water table or zone of saturation. The shallow hydrologic environment of wetlands creates unique biogeochemical conditions that distinguish it from aquatic and terrestrial environments. Wetlands are a fundamental hydrologic landscape unit that generally forms on flat areas or shallow slopes, where perennial water lies at or near the land surface, either above or below. Wetlands tend to form where surface and ground water accumulate within topographic depressions. Wetland Hydrology deals with the general hydrologic properties that make wetlands unique, and to provide an overview of the processes that control wetland hydrologic behavior.

Tunnel Hydrology

All over the globe, the response of the tunnel discharge to rainfall was rapid and indicated that the flow had largely by-passed the soil matrix. Therefore, rainfall and surface runoff were infiltrating down to the subsoil (B2) horizon through established tunnel pathways.

Hydrogeology

Hydrogeology is the study of groundwater – it is sometimes referred to as geohydrology or groundwater hydrology. Hydrogeology deals with how water gets into the ground (recharge), how it flows in the subsurface (through aquifers) and how groundwater interacts with the surrounding soil and rock (the geology).

Importance of Hydrogeology

- Though many parts of the globe have abundant sources of water and adequate rainfall, there are still places where the only source is underground water.

- In such places, hydrogeologists are bound to trace appropriate reserves and evaluate the quantity of water that can be hauled out without causing any damage to the sub-terranean aquifers or the ecosystems.
- By knowing the movement of underground water, hydrogeologists can also predict the hazards of pollution and ways to protect these resources.
- Without the know-how of hydrogeology, there is no way we can forecast the condition of the water sources. Thus, hydrogeology is a very important aspect in the protection of the most necessary natural resource.

Hydrogeologists apply this knowledge to many practical uses. They might:

- Design and construct water wells for drinking water supply, irrigation schemes and other purposes.
- Try to discover how much water is available to sustain water supplies so that these do not adversely affect the environment – for example, by depleting natural base flows to rivers and important wetland ecosystems.
- Investigate the quality of the water to ensure that it is fit for its intended use.
- Where the groundwater is polluted, they design schemes to try and clean up this pollution.
- Design construction dewatering schemes and deals with groundwater problems associated with mining.
- Help to harness geothermal energy through groundwater-based heat pumps.

Hydrochemistry

Hydrochemistry is the science of the chemical composition of natural waters and the laws governing the changes in composition as a result of the chemical, physical, and biological processes occurring in the surrounding environment. Since it is the science of the chemistry of the hydrosphere, hydrochemistry is part of both geochemistry and hydrology.

Hydrochemistry is important for the development of a number of related sciences, including petrology, mineralogy, soil science, hydrogeology, and hydrobiology. Knowledge of the chemical composition of water (which determines its quality) is necessary for such areas of practical activity as water supply, irrigation, and fish culture. Hydrochemical data are important in evaluating the corrosion of construction materials (concrete and metal), in analyzing mineral waters, in mineral prospecting (petroleum, ore deposits, and radioactive substances), and the like. The study of the chemical composition of water becomes particularly important when combatting the pollution of water basins by waste waters. Three subdivisions of hydrochemistry can be distinguished in its present stage of development:

1. The first subdivision deals with the formation of the chemical composition of natural waters. It includes the study of water as a solvent for the complex of minerals of the earth's crust and the study of the chemical processes that occur in water as it interacts with rocks, soils, organisms, and

the atmosphere. Investigations are also carried out on the solubility of substances found in nature, their state in solution and their stability, as well as on the sorption, exchange, and oxidation-reduction processes. In this subdivision, closely related to geochemistry, we must include the general questions of the cycle of substances and the migration of elements in the hydrosphere.

2. The second subdivision deals with the chemical composition and hydrochemical cycle of certain forms of natural water and the dependence of their changes on the physicogeographical conditions of the surrounding environment. This broad subdivision borders closely on hydrology, and it is divided into the chemistry of rivers and lakes, chemistry of the sea, and the chemistry of underground water and atmospheric water.

The chemistry of surface waters studies the chemical composition of water in rivers, lakes, and artificial reservoirs, changes in the composition with land or water area and with depth, seasonal diurnal variations, and the conditions of the determination of its composition in its dependence on the surrounding environment. It is of great importance to predict the chemical composition of water reservoirs built in arid regions and to combat pollution in reservoirs. Investigations of salt lakes, rich in minerals, are very important for the chemical industry.

The chemistry of the sea is closely related to oceanography. In addition to the study of the dependence salinity, biogenic substances, and dissolved gases on hydrodynamic, hydrometeorological, and hydrobiological factors, it also studies the form and content of trace elements, the origin of organic substances, the processes of the metamorphization of organic substances, and the processes of the interaction of seawater with river and sea bottom sediments.

The chemistry of underground waters includes the study of the chemical composition of ground, stratal, artesian, and mineral waters and of the waters from petroleum deposits. Particularly stressed are the formation of the composition of waters and the processes of the interaction of water with the surrounding rocks, which take place at high pressures and often at elevated temperatures with retarded water exchange and characteristic microbiological conditions. For a long time, the study of mineral waters, quite varied in composition and origin, has been of great importance.

3. The third subdivision consists of the methods used in hydrochemical investigations. This subdivision is a special branch of analytical chemistry, as applied to the characteristics of the analysis of natural waters. At present, fiydro-chemistry makes extensive use of spectroscopy, chromatography, polarography, tagged atoms, and other physico-chemical methods. A large subdivision of analysis is the determination of the components of polluted water.

Ecohydrology

Ecohydrology (EH) a subdiscipline of hydrology that focuses on ecological processes occurring within the hydrological cycle and strives to utilize such processes for enhancing environmental sustainability. Has been developed under the International Hydrologcal Programme of UNESCO.

The general assumption of EH is to reverse degradation and achieve sustainable water and ecosystems in anthropogenically modified basins. In addition to the reduction of erosion, nutrients, and pollutant emissions there is a necessity to regulate ecological processes based on understanding "water – biota interactions," from molecular (e.g., microbial loop) to ecosystem (biomanipulation) and to landscape scales (reforestation, creation land/water ecotone zones).

Vadose Zone Hydrology

Vadose zone hydrology is also termed the unsaturated zone is the part of Earth between the land surface and the top of the phreatic zone i.e. the position at which the groundwater (the water in the soil's pores) is at atmospheric pressure ("vadose" is from the Latin for "shallow"). Hence the vadose zone extends from the top of the ground surface to the water table. Water in the vadose zone has a pressure head less than atmos pheric pressure, and is retained by a combination of adhesion (funiculary groundwater), and capillary action (capillary groundwater). If the vadose zone envelops soil, the water contained therein is termed soil moisture. In fine grained soils, capillary action can cause the pores of the soil to be fully saturated above the water table at a pressure less than atmospheric. In such soils, therefore, the unsaturated zone is the upper section of the vadose zone and not identical to it. Movement of water within the vadose zone is studied within soil physics and hydrology, particularly hydrogeology, and is of importance to agriculture, contaminant transport, and flood control. The Richards equation is often used to mathematically describe the flow of water, which is based partially on Darcy's law. Groundwater recharge, which is an important process that refills aquifers, generally occurs through the vadose zone from precipitation.

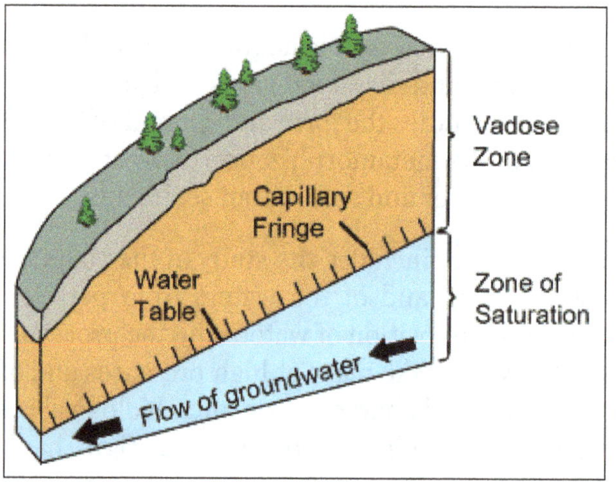

Moisture State in the Vadose Zone

The water content and matric pressure within the vadose zone influence and in turn are influenced by conditions in the saturated zone and atmosphere. The distribution of vadose water at a given time also depends on the energy state (whose components include matric and gravitational potential), wetting/drying history, and dynamics of the water itself. If there is no flow, one can infer that the gradient of total potential is zero, so if the matric and gravitational components are the only significant ones, they add to a constant total potential. Figure (a) shows this type of hydrostatic profile for the case where a water table is present. Since the matric pressure in this case is linear with depth and θ is controlled by the water retention properties of the medium, for a uniform vadose zone, the θ profile (not shown here) mimics the shape of the water retention curve. Given time, approximate versions of such a hydrostatic profile may develop in portions of a profile where water movement is negligible. If water flows vertically downward at a steady rate in a homogeneous medium, the total gradient must be constant, but the matric pressure does not cancel out the gravitational potential, as illustrated in figure (b). In the general case of unsteady flow, the matric pressure profile cannot be determined so simply, and may take on an irregular form as in figure (c).

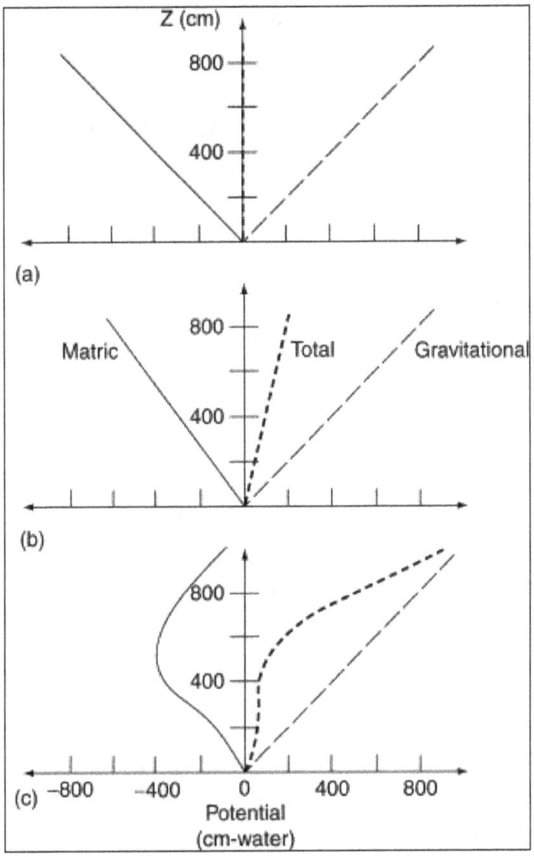

Profiles of matric, gravitational, and total potential for idealized situations of (a) static water, (b) steady downward flow, and (c) unsteady flow. The water table is at z = 0.

The uppermost part of the water distribution profile is sometimes described in relation to field capacity, defined as the water content of a soil profile when the rate of downward flow has become negligible 2 or 3 days after a major infiltration. This concept is used in agriculture to indicate the wettest soil conditions to be considered for plant growth, and sometimes is mentioned in hydrologic investigations related to soil moisture storage. Field capacity is implicitly associated with the entire soil profile through the root zone, including preferential-flow characteristics and, especially, flow-retarding layers that enable layers above them to retain high water content.

In a portion of the vadose zone immediately above the water table, it may happen that all pores are filled with water, held by capillary forces. The depth interval that is saturated but above the water table is called a capillary fringe. In a hydrostatic profile, this corresponds to a flat portion of the retention curve between saturation and an air-entry pressure. Some media do not have a significant capillary fringe because their retention characteristics have the air-entry pressure at essentially zero. Where the water table fluctuates, the hydrostatic equilibrium needed for a capillary fringe may take considerable time to establish. Soil–water hysteresis would make for a different capillary fringe with a falling water table than with a rising water table.

Isotope Hydrology

Isotope hydrology is the study of the isotopic signatures of water. This subfield of hydrology utilizes isotopic dating to determine the origin and age of water throughout its movement within the hydrologic cycle. Isotopic dating involves measuring the levels of deviation in the isotopes of oxygen and hydrogen in water. Researchers are able to determine groundwater dated as far back as the Ice Age by using these techniques. Isotope hydrology deals with water usage policy, mapping aquifers, conservation of water resources, and maintaining pollution levels.

Forest Hydrology

Forest hydrology combines aspects of two separate disciplines: hydrology and forestry. Hydrology is the science that studies the waters of Earth. Hydrology seeks to understand where water occurs; how water circulates; how and why water distribution changes over time; the chemical and physical properties of water; and the relation of water to living organisms.

Forestry is the science that seeks to understand the nature of forests and the interactions between the parts comprising a forest. Forest management started in the 1700s as a form of large-scale farming to improve yields of timber and fiber from forests. In the United States, watershed protection has been an integral part of forest management since its origins. The Organic Administrative Act of 1897 stated that forest reserves were to protect and enhance water supplies, reduce flooding, secure favorable conditions of water flow, protect the forest from fires, and provide a continuous supply of timber. The 1911 Weeks Act authorized the acquisition of federal lands in the East for the express purpose of protecting the watersheds of navigable waterways.

In recent decades, forestry has adopted more of an ecosystem management approach while still including timber production as an important goal. Although a forest is an ecosystem dominated by trees, a healthy forest includes other plants as well as soil, terrestrial and aquatic animals, and water—plus people who use the forest and its resources. Modern forest management therefore requires not only an understanding of forest science, soil science, and hydrology, but also principles of wildlife biology, land-use planning, and recreation planning.

Water and Forested Ecosystems

Ecologists consider water to be the defining part in an ecosystem, including the forest ecosystem. Water shapes the physical landscape through erosion and deposition. It also shapes the biological parts of the ecosystem by its presence or absence; its quantity and quality; and its occurrence and distribution. The water cycle plays a key role in ecosystem functions and processes.

Forests, in turn, are vital to the water cycle and to water quality. In essence, the forest acts like a giant sponge, filtering and recycling water. Approximately 80 percent of U.S. fresh-water resources are estimated to originate in forests, which cover one-third of the U.S. land area.

Tree leaves intercept water from rain, snow, and fog; the leaves also release water back to the atmosphere by evapotranspiration. Tree roots extract water from the soil while helping hold the soil in place. Forested land reduces the surface impact of falling rain through interception and delay of water reaching the surface. Forestland also decreases the amount and velocity of storm runoff over the land surface. This in turn increases the amount of water that soaks into the ground, a portion of which can ultimately recharge underlying aquifers. Conversely, water from hydraulically connected surficial aquifers may enter streams and wetlands, helping to maintain their water levels during dry periods.

Undulating forest surrounds meadows, lakes, and rivers of the Charlevoix Conservation Area, a coastal watershed in Quebec, Canada. This region, which illustrates a high-quality forested watershed, has been designated a biosphere by the United Nations Educational, Scientific and Cultural Organization (UNESCO).

Forests and the Hydrologic Cycle

The surface water in a stream, lake, or wetland is most commonly precipitation that has run off the land or flowed through top soils to subsequently enter the water body. If a surficial aquifer is present and hydraulically connected to a surface-water body, the aquifer can sustain surface flow by releasing water to it.

In general, a heavy rainfall causes a temporary and relatively rapid increase in streamflow due to surface runoff. This increased flow is followed by a relatively slow decline back to baseflow, which is the amount of streamflow derived largely or entirely from groundwater. During long dry spells, streams with a baseflow component will keep flowing, whereas streams relying totally on precipitation will cease flowing.

Generally speaking, a natural, expansive forest environment can enhance and sustain relationships in the water cycle because there are less human modifications to interfere with its components. A forested watershed helps moderate storm flows by increasing infiltration and reducing overland runoff. Further, a forest helps sustain streamflow by reducing evaporation (e.g., owing to slightly lower temperatures in shaded areas). Forests can help increase recharge to aquifers by allowing more precipitation to infiltrate the soil, as opposed to rapidly running off the land to a downslope area.

Riparian Areas

The riparian zone is broadly defined as the area between a body of water and the upland parts of the landscape that are rarely flooded except under the most extreme conditions. But the term also can refer more specifically to the immediate streamside area.

Riparian areas represent less than 10 percent of most forest ecosystems, yet these areas often are the most productive portions. Compared to upland regions, riparian areas have more water available; the vegetation is more robust; the soils are deeper; the timber often is of higher quality; and the water bodies have more shade. The riparian zone also may include wetlands bordering streams and lakes. This combination of factors makes riparian areas among the most heavily used portions of a forest. Riparian and wetland areas provide abundant and reliable forage for wildlife, as well as transportation corridors. They also may receive heavy human use for recreation.

Riparian zones also are attractive destinations for logging and for livestock grazing; as a result, riparian areas in forests are sometimes heavily damaged, especially in the forests of the arid American Southwest. Fortunately, riparian areas respond well to good management practices.

Aquatic Biodiversity

Forest lands and waters are vitally important in maintaining biodiversity and providing habitat for fish and wildlife, including threatened or endangered aquatic species. In the United States, over one-third of national forest lands are critical for maintaining aquatic biodiversity and protection of listed species.

For aquatic species, watersheds provide the basic unit of any conservation strategy. Many watersheds also contain isolated habitats with unique characteristics producing a high potential for rare species. Some species occur only near a single spring or in a single stream within a given watershed. Lands set aside to protect these unique habitats also benefit the entire watershed and its ecosystem.

Forest Management and Watershed Quality

Wind, fire, insects, and disease are all part of properly functioning, healthy ecosystems in watersheds. For example, natural fires, although temporarily devastating, periodically restore the balance between vegetation types, and release nutrients from the vegetation and soil. In contrast, widespread clear-cut logging and excessive or improper road-building can degrade watersheds, as can land uses such as ski runs and housing projects. Many human activities can increase overland runoff, resulting in more erosion of the land surface and concurrently reducing the amount of water that soaks in the ground to potentially reach nearby streams or recharge underlying aquifers.

Rill and gully erosion caused by flowing water is evident on this steep slope following a forest fire. The soil loss can be substantial, and denuded slopes can be difficult to revegetate.

Moreover, fire prevention and suppression have created "imbalanced" forests with excessive amounts of undergrowth and dead vegetative matter that serve as fuels when fire does occur. Hence, these forests are at increased risk of high-intensity, destructive fires.

Watershed management and restoration may include controlled thinning, prescribed burning, and other management practices to restore the proper balance of timber, undergrowth, and grassy meadows in the watershed. Restoration also may include planting of appropriate native plants.

Effects of Roads

Improperly engineered roads in forests can increase erosion and significantly increase the risk of landslides. Both of the adverse effects are more severe when roads are numerous, and when they either cross or run parallel to streams. For example, heavy precipitation in Oregon and Washington during the mid-1990s resulted in many landslides, with a correlation between the slides and the frequency and density of roads. With respect to erosion and sedimentation, water runoff flowing along and across roads picks up sediment, which can then be deposited in nearby lakes and streams. This siltation can degrade or destroy habitat for aquatic organisms that require clear water and silt-free benthic (bottom) substrates.

Effects of Fire

Destructive fires that remove large amounts of organic matter in a forest cause loss of nutrients from the soil as the detrital cover (i.e., dead and decaying materials on the forest floor) and upper soil layers are burned and eroded. Moreover, fires can adversely affect the quality of streams and lakes in the burned region as well as tributary watersheds downstream. Surface runoff that otherwise would have been slowed or absorbed by living and dead vegetative matter on the forest floor now runs unimpeded down bare (or nearly bare) slopes. The increased velocity carries more soil particles and loose vegetative matter downslope, along with any adsorbed nutrients.

Darcy's Law

Darcy's law is the most fundamental law in groundwater science. It was proposed by Henry Darcy. It is one of the first mathematical relationships proposed for water flow through saturated porous media. The simplicity of the mathematical expression in Darcy's law facilitates its widespread use in quantification of groundwater flow. It will not be an exaggeration if we say that this law forms the basis of all quantitative hydrogeology. The law gives general relationship for groundwater flow through saturated porous media.

In the figure there is an inclined main cylindrical tube filled with porous media. The water is put in the system through water inlet on left and it comes out from the water outlet on the right. The water in the main cylindrical tube flows from left to right as shown by arrow. There are two vertical small diameter cylinders (manometers) near water inlet and outlet. They are meant for the purpose of measuring head of the water flowing through the porous media (manifested by elevation of water in the tubes) at the entry and exit positions in the given experimental setup.

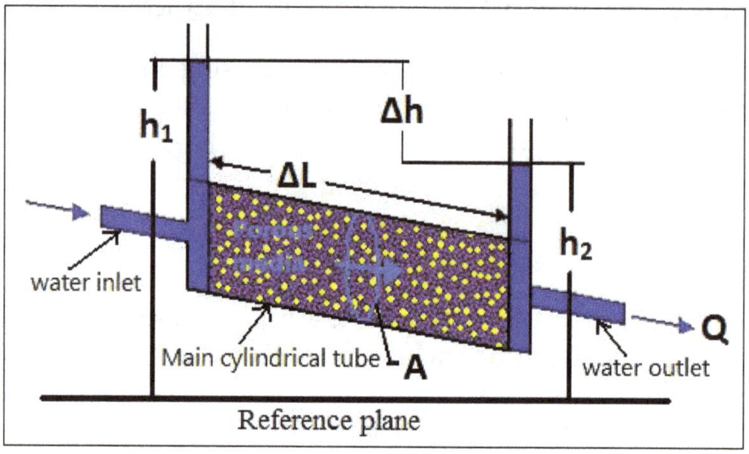

Experimental setup for water flow through saturated porous media.

Further,

- h_1 is the initial head of water with respect to a reference plane (Unit: meter).
- h_2 is the final head of water with respect to a reference plane (Unit: meter).
- Δh is the difference between final and initial head ($h_2 - h_1$) (Unit: meter).
- ΔL is the length of flow path (Unit: meter).
- A is the cross sectional area through which water flow takes place (Unit: meter²).
- Q is flow rate of water flowing that is the volume of water discharged at the end of the cylindrical tube in a given time 't' (Unit: meter³/day).

The flux of water (v) through the system shown in figure is volume rate of water flow through unit cross section area; alternatively volumetric flow per unit area. It can be mathematically shown by:

$$v = \frac{Q}{A}$$

Darcy's observation elucidated with the experimental setup shown in figure was that the flux of water (v) through a saturated porous media for a given length (ΔL), under influence of a given head loss (Δh) is:

- Directly proportional to the head loss (Δh).
- Inversely proportional to the length of the flow path (ΔL).

This can be mathematically expressed as:

$$v \propto \frac{\Delta h}{\Delta L}$$

The ratio of head loss (Δh) to the corresponding length (ΔL) shown as $\frac{\Delta h}{\Delta L}$ in previous equation is referred to as hydraulic gradient.

If we formulate a statement of Darcy's law based on empirical observations, it would read as: "the flux of water (v) through saturated porous media is directly proportional to the hydraulic gradient".

Further replacing the proportionality sign by equality constant K gives Darcy's law shown as:

$$v = -K\frac{\Delta h}{\Delta L}$$

Here the constant K has unit of meters/day and is referred to as coefficient of permeability, more popularly known as hydraulic conductivity. The minus sign in the above equation have been introduced by the fact that in the given experimental set up Δh will always be negative. This is explained below:

Since $\Delta h = h_2 - h_1$ with a condition that always $h_2 < h_1$; hence Δh is always negative. This also points to the fact that water flow through porous media is always in direction of decreasing water head. It implies that as water moves through a porous media it loses energy to other forms. This could be attributed to the resistance offered by the media to the flowing fluid. In figure, it can be observed that head varies along water flow path; that is Δh varies with Δl. Thus the Darcy's law in differential form cane be written as:

$$v = -K\frac{dh}{dL}$$

As $v = \frac{Q}{A}$ the previous equation is popularly also written as:

$$Q = -KA\frac{dh}{dL}$$

Groundwater flow is through porous saturated media and Darcy's law gives an equation for water flow through porous saturated media. Hence by this analogy, Darcy's law has been used as most fundamental equation for quantification of groundwater flow.

Groundwater Velocity in Context of Darcy's Equation

If we examine the water flux (v) in the experimental setup above, it turns out to be velocity of water through the main cylindrical tube; with unit meter/day. Hence this water flux (v) is also referred to as specific discharge or Darcy velocity. Thus it is expected that this value should give an approximate estimate of the velocity with which groundwater will flow through different media.

However, a subsequent scrutiny revealed that Darcy velocity is an under estimate. In this equation we observe that denominator (A) is the area of cross section of the main cylindrical tube shown in figure. However if we visualize water flow through porous media in the main cylindrical tube of figure, it turn out that actual water flow takes place through pores/voids and not the whole cross section as given in equation $v = \frac{Q}{A}$. This is schematically shown with figure, where a cross section of the main cylindrical tube shows that water moves through so many pores/voids (shown as small circles) and not the whole cross-section 'A' of the cylindrical tube.

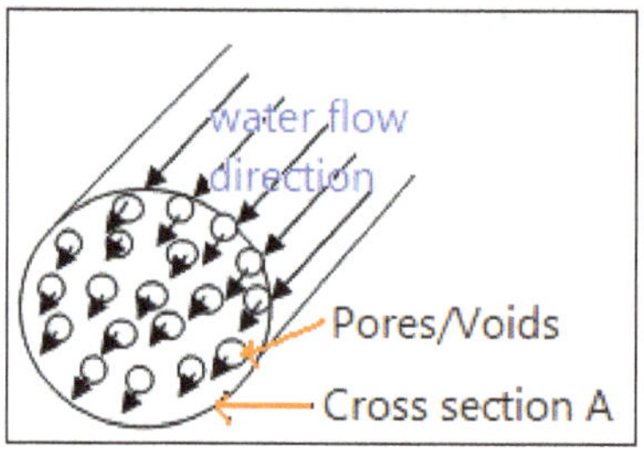

A schematic drawing showing water flow through pores/voids.

In this case it becomes obvious that in equation $v = \frac{Q}{A}$ by putting 'A' in denominator, we are putting a higher value. Thus the estimated water flow velocity through porous media is an underestimate. In order to rectify this anomaly, a solution is given that if the cross section area (A) of the main cylindrical tube is multiplied with porosity, which is an estimate of void volume, the resultant velocity would approximate the real water velocity. In order to make it distinct, this resultant velocity has been given names like 'pore water velocity/seepage velocity/average interstitial velocity etc.' For our convenience we will use 'vp' to denote this velocity. This velocity is mathematically represented by:

$$v_p = \frac{Q}{A * p}$$

$$v_p = \frac{v}{p} \text{ as } v = \frac{Q}{A}$$

Where Q and A are same as mentioned earlier, while 'p' represents porosity of the media. Since porosity is always less than one; multiplying 'A' in denominator by porosity will reduce the denominator. This will give a relationship shown by:

$$v_p > v$$

A further insight revealed that all pores in a porous media are not interconnected and does not contribute to linear flow of water through the media. It is the interconnected pores which facilitates linear flow of water through a porous media. An understanding of the fact that linear water flow which is ultimately discharged (like at the right end of main cylindrical tube in figure) after flowing through a porous media is the water which has flown through interconnected pores measured by effective porosity. The linear water flow through a porous media is effectively the flow through interconnected pores and not through all the pores in the media. Thus 'pore water velocity/seepage velocity/actual velocity through pores etc.' (vp) estimated by incorporating porosity in equation $v_p = \frac{Q}{A * p}$ and $v_p = \frac{v}{p}$ is still not the real estimate of water flow velocity through porous media. Now in order to rectify this, a solution is that if the cross section area (A) of the main cylindrical tube given in equation $v = \frac{Q}{A}$ is multiplied with effective porosity (e), which is an estimate of interconnected void, the resultant velocity would approximate the real water velocity. Thus we have a new velocity term 'average linear velocity' denoted here for our convenience by 'vl'. This velocity is mathematically represented by:

$$v_l = \frac{Q}{A * e}$$

$$v_l = \frac{v}{e} \text{ as } v = \frac{Q}{A}$$

Where, Q and A are same as mentioned earlier, while 'e' represents effective porosity of the media. Since effective porosity is always less than porosity and both of them are less than one; multiplying 'A' in denominator by effective porosity will further reduce the denominator. This will give a relationship shown by:

$$v_l > v_p > v$$

The relationship shown in equation-10 is further explained with an example.

Example: It is given that area of cross section (A) of a porous media/formation through which water flows is 10 m², porosity (p) of the formation is 0.5 and effective porosity (e) is 0.25. We will estimate the different velocities at a given discharge (Q) of 100 m³/day.

As per equation $v = \frac{Q}{A}$, Darcy velocity $v = \frac{Q}{A} = \frac{100}{10} = 10 \text{ m/day}$.

As per equation $v_p = \frac{v}{p}$, seepage velocity $v_p = \frac{v}{p} = \frac{10}{0.5} = 20 \text{ m/day}$.

As per equation $v_l = \frac{v}{e}$, average linear velocity $v_l = \frac{v}{e} = \frac{10}{0.25} = 40 \text{ m/day}$.

Thus, $v_l > v_p > v$.

Limitation of Darcy's law

Darcy law is valid for laminar flow. Reynold's number (Re) is used to distinguish laminar flow from turbulent flow as given by equation:

$$Re = \frac{p * v * d}{\mu}$$

Where, ρ is density of fluid, v is magnitude of its velocity, d is boundary to fluid flow approximating to mean pore diameter, μ is viscosity of the fluid.

It has been observed that Darcy's law is valid for Nr < 1 and may be up to Nr = 10. However most groundwater flow under natural condition occurs with Nr<1 and hence Darcy's law is valid. It is also likely that non-linear diffused flow through very fine grained particles will not be governed by Darcy's law. Shekhar estimated Reynold's number in vicinity of tube wells discharging water at very high discharge and found it to be much higher than the critical limit set for onset of turbulence in flow regime. In such cases where flow is not laminar; use of Darcy's law for quantification of groundwater flow should be avoided. It sets limitation to use of Darcy's law for quantifying such groundwater flows where turbulence has been induced by very high hydraulic gradient. Groundwater flow through sloping large cracks and opening is also likely to be turbulent in nature and Darcy's law may not be valid.

Application of Darcy's Law

Darcy's law is the most fundamental equation describing water flow in saturated porous media and almost all other equations in hydrogeology are based on Darcy's law. Some of the important equations which are based on the Darcy's law are:

- Equation for three dimensional groundwater flow in saturated confined and unconfined aquifer.
- Equation for steady state flow in confined and unconfined aquifers.
- Darcy's law is widely used for basic quantification of any groundwater system.

Besides this, Darcy's law forms the basis for quantification of groundwater system using flow nets and similarly much other estimation.

References

- What-is-hydrogeology: iah.org, Retrieved 07, August 2020
- Forest-hydrology: waterencyclopedia.com, Retrieved 26, May 2020
- Hydrology-and-its-branches-315484240: researchgate.net, Retrieved 02, April 2020
- Vadose-zone, agricultural-and-biological-sciences: sciencedirect.com, Retrieved 25, June 2020
- Hydrogeology-defined, hydraulics-civil-engineering-96784: brighthubengineering.com, Retrieved 11, Feb 2020

2

The Hydrologic Cycle

The cycle that explains the continuous circulation of water from the time it falls to the ground until it returns to the atmosphere is known as the hydrologic cycle. This cycle also involves the energy exchange which leads to temperature changes. This chapter has been carefully written to provide an easy understanding of the varied facets of water cycle.

Earth contains enormous amounts of water in the form of reservoirs. Water exists in the atmosphere, lithosphere, hydrosphere and biosphere. Among all these segments, water masses are in continuous circulation. It is, in general, called as the hydrologic cycle. It is also called as the World's Great Water Cycle, since it is the driving wheel for all the movements of available water resources on the planet earth. It is necessary to identify the cyclic and circulatory routes of water masses in all the spheres of the earth. Several processes and factors are involved in driving the global water circulation.

The planet earth is called as the water planet. It is due to the fact that the earth is probably the only planet, in the solar system, which has a huge mass of water. On the planet earth, the primitive life got originated only from the water mass. Most of the earth's surface is covered with water. Water exists in all the segments of the earth. The hydrosphere is the sphere of water on earth. It is a discontinuous layer containing both fresh water and saline water.

The world's water is not in static from. It is in continuous motion. It also transforms from one state to the other. The water gets temporarily stored and moved in the form of surface water in rivers, lakes, ponds, ice caps, and below the surface as groundwater. Oceans form the biggest and largest water reservoirs. All these water bodies are called as water reservoirs. The duration of stay and storage of water in every reservoir varies due to varying geological, environmental and other conditions. This is called as the residence time of water. Water moves from one reservoir to the other. The sun's radiant energy plays a very significant role in this movement.

Atmospheric pressure, mind blows, temperature and the amount of water vapour present in the air, all play a dominant role in these processes. The rates of movement of water and the quantities involved the cyclic processes are the major aspects involved in the hydrological sciences. There is an endless circulation of water among all the spheres of the earth. It is popularly known as the hydrologic cycle. It is necessary to learn about the hydrologic cycle, when we intend analyse the water resources of the region and the world.

Water gets transformed from liquid to solid, solid to liquid, liquid to vapour, vapour to liquid and vapour to solid states. The sun's radiation, acceleration due to gravity, ability of the water to flow and several other properties of water, make this transformation more effective and regular. The basic input to the world's water masses comes from precipitation.

Precipitated rain (or) snow falls overland. Processes like infiltration and peroration moves the water down to the groundwater systems. Some amount of water flows towards the sea as runoff.

The surface water collected in lakes, ponds, swamps, seas and oceans get evaporated into the atmosphere. The vegetation transpires the water collected from the soil moisture. Evaporated and transported water enters into the atmosphere as vapour. Collected water vapour gets condensed to form the clouds.

Clouds more towards the land and starts precipitating again. These processes continue. This endless circulation of water is known as the hydrologic cycle. It is of two kinds as terrestrial hydrologic cycle and global hydrologic cycle. The terrestrial hydrological cycle is of a special interest as the mechanism of formation of water resources on a given area of the land, like a river basin or a watershed. The global hydrological cycle deals with its role on the global climate and other geological and physical processes. It is obvious that the role of different processes involved in the hydrological cycle and their description have to depend on the chosen spatial- temporal scales.

Scales for Study of Hydrologic Cycle

From the point of view of hydrologic studies, two scales are readily distinct. These are the global scale and the catchment scale.

Global Scale

From a global perspective, the hydrologic cycle can be considered to be comprised of three major systems; the oceans, the atmosphere, and the landsphere. Precipitation, runoff and evaporation are the principal processes that transmit water from one system to the other. This illustration depicts a global geophysical view of the hydrologic cycle and shows the interactions between the earth (lithosphere), the oceans (hydrosphere), and the atmosphere. The study at the global scale is necessary to understand the global fluxes and global circulation patterns. The results of these studies form important inputs to water resources planning for a national, regional water resources assessment, weather forecasting, and study of climate changes. These results may also form the boundary conditions of small-scale models/applications.

Catchment Scale

While studying the hydrologic cycle on a catchment scale, the spatial coverage can range from a few square km to thousands of square km. The time scale could be a storm lasting for a few hours to a study spanning many years. When the water movement of the earth system is considered, three systems can be recognized: the land (surface) system, the subsurface system, and the aquifer (or geologic) system. When the attention is focused on the hydrologic cycle of the land system, the dominant processes are precipitation, evapotranspiration, infiltration, and surface runoff. The land system itself comprises of three subsystems: vegetation subsystem, structural subsystem and soil subsystem. These subsystems subtract water from precipitation through interception, depression and detention storage. This water is either lost to the atmospheric system or enters subsurface system. The exchange of water among these subsystems takes place through the processes of infiltration, exfiltration, percolation, and capillary rise.

Figure shows the schematic of the hydrologic cycle at global scale, in the earth system, and micro-scale view of the cycle in the land system. Figure gives a schematic presentation of the hydrologic cycle of the earth system. Detailed schematic of the hydrologic cycle in the land system is shown in figure.

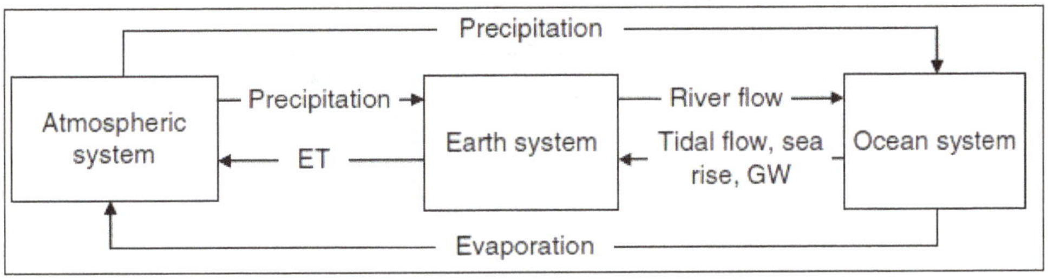

A global schematic of the hydrologic cycle.

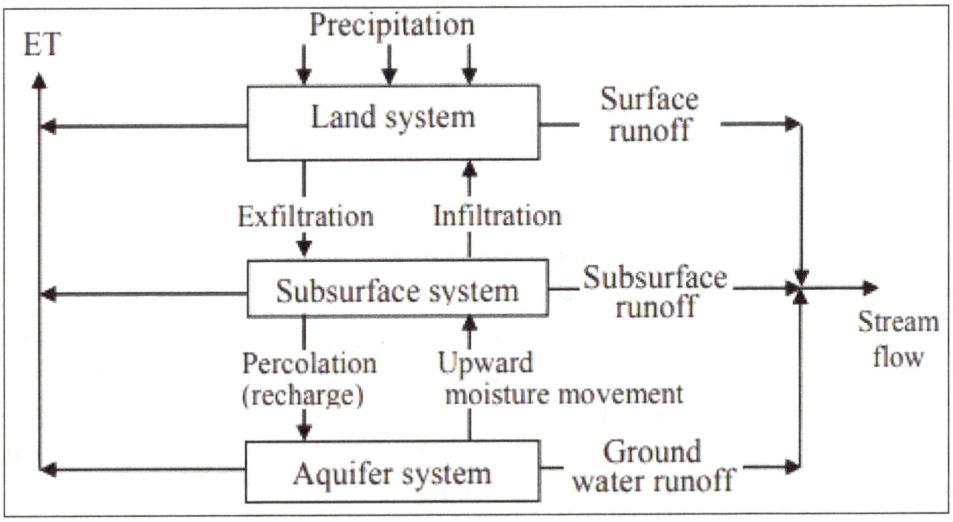

A schematic of the hydrologic cycle of the earth system.

Time Scales in Hydrologic Cycle

The time required for the movement of water through various components of the hydrologic cycle varies considerably. The velocity of stream flow is much higher compared to the velocity of ground water. The time-step size for an analysis depends upon the purpose of study, the availability of data, and how detailed the study is. The estimated periods of renewal of water resources in water bodies on the earth is given in table. The time step should be sufficiently small so that the variations in the processes can be captured in sufficient detail but at the same time, it should not put undue burden on data collection and computational efforts.

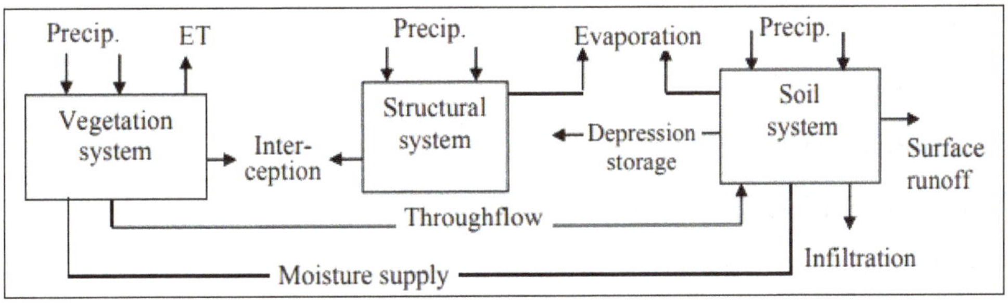

A detailed schematic of the hydrologic cycle in the land system.

Table: Periods of water resources renewal on the Earth.

Water of hydrosphere	Period of renewal
World ocean	2500 years
Ground water	1400 years
Polar ice	9700 years
Mountain glaciers	1600 years
Ground ice of the permafrost zone	10000 years
Lakes	17 years
Bogs	5 years
Soil moisture	1 year
Channel network	16 days
Atmospheric moisture	8 days
Biological water	Several hours

Global Hydrologic Cycle

The global hydrological cycle describes the transport and occurrence of water in all three phases and its transformations from one phase to the other, on a global scale. Most of the Earth's water masses reside in the oceans. Continental water makes up about 3.5 percent of the Earth's water. About three-quarters of this amount (29 million cubic kilometers) is present as polar ice caps and glaciers. About 5.3 million cubic kilometers as deep groundwater. Thus, only the remaining fraction can take part in the water exchange between the oceans, the atmosphere, and the continents. This remaining part includes shallow groundwater and soil moisture, water in lakes, reservoirs, and swamps, water storage in river channels, biosphere water. The amount in the atmosphere is only 0.013 million cubic kilometres. But it plays a very crucial role in the global water cycle.

Components of the Global Hydrological Cycle

The hydrological cycle consists of a collection of reservoirs (each having a particular mass of water substance) and movement (fluxes, measured in units of mass or volume per unit time) of water substance between these reservoirs. The following are reservoirs of the global hydrological cycle:

- Global oceans.
- Ice masses:
    - Antarctic ice sheet,
    - Greenland ice sheet,
    - Mountain glaciers,
    - Arctic ice,
    - Sea ice.

- Continental seasonal snow.
- Surface fresh water:
    - Lakes,
    - Rivers,
    - Marshes and wetlands.
- Subsurface water:
    - Soil moisture,
    - Permafrost,
    - Ground water:
        - Deep aquifers,
        - Volcanic water.
- Biospheric water.
- Atmospheric water vapor.
- Clouds:
    - Liquid,
    - Ice.

Figure gives the relative sizes of the land, ocean, and atmospheric reservoirs of the global hydrological cycle and the processes (evaporation, precipitation, and runoff) that lead to interchanges among these reservoirs. The ocean is obviously the largest reservoir, followed (by a factor 40 smaller) by land, and followed by an additional factor of 2500 by the atmosphere.

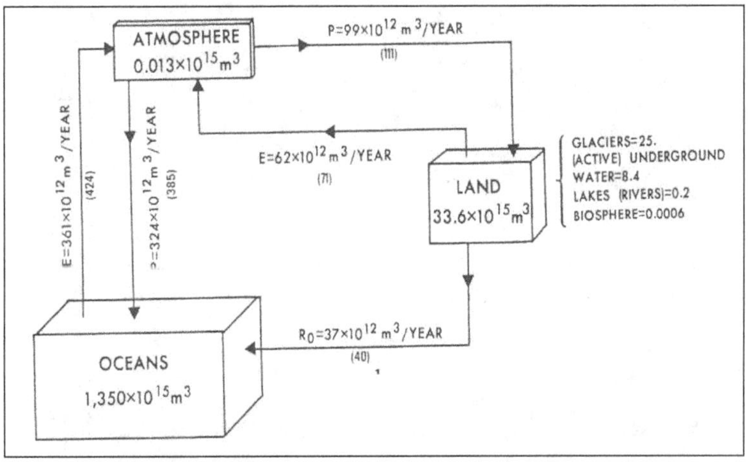

The amounts of water stored in the oceans, land, and atmosphere, and the amounts exchanged annually between the different reservoirs through evaporation, precipitation, and runoff.

Precipitation over land is larger than evaporation, meaning that considerable water evaporated over oceans eventually falls on land and returns to the ocean by way of runoff through rivers.

By dividing the reservoir volume by its loss rate, we get an estimate of the residence time or timescale of H_2O in each reservoir. Table describes the largest reservoirs and the typical residence times, which range from thousands of years for oceans to about a week for the biosphere.

Table: Reservoirs of Available Water (liquid equivalent) on Earth.

Reservoir	Volume (10^6 km^3)	Percent of Total	Residence Time
Oceans	1350	97.3	$10^3 - 10^4$ year
Glaciers	29	2.1	$10^1 - 10^3$ year
Aquifers	8	0.6	2 weeks — 10^4 year
Lakes	0.1	0.01	10 year
Soil Moisture	0.1	0.01	52 days
Atmosphere	0.013	0.001	10 days
Rivers	0.002	0.0002	2 weeks
Biosphere	0.001	0.0001	6 days

Role of the Global Hydrological Cycle

Global Oceans

Global oceans participate in the global energy balance by transporting heat energy, which is preferentially received in the tropics by solar radiation, toward the Polar Regions. Over half of the solar energy reaching the earth is first absorbed by oceans and then transmitted to other regions. Global oceans take excess CO_2 from the atmosphere at a rate that depends on ocean temperature. The partial pressure of CO_2 in cold water is depressed by enhanced solubility and (in the spring bloom) phytoplankton growth, making cold water a sink for atmospheric CO_2.

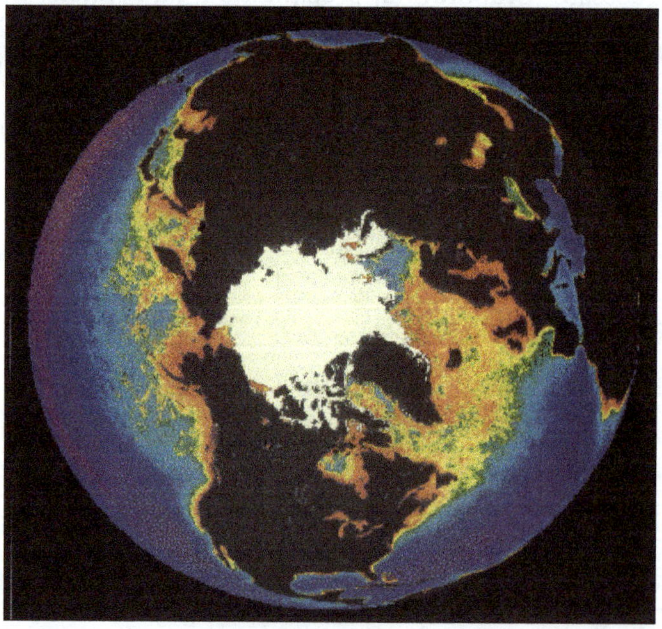

These surface water biological processes support a full spectrum of the food chain from phytoplankton to whales. Global oceans play a significant role in determining where and how much water vapor enters the atmosphere, which in turn influences the global energy balance. The massiveness of the global oceans also introduces a thermal inertia into the climate system: ocean processes are slow in comparison to atmospheric processes, so changes in ocean structure can introduce slow changes of climate. This fact has become more evident to us in recent years as we have come to understand the El Nino/Southern Oscillation (Climate Variability) which we now know causes perturbations of weather in certain regions for several successive months.

Ice Masses

Ice masses offer storage of H_2O on time scales ranging from a few months for continental seasonal snow pack to thousands of years for the Antarctic ice mass. They all contribute to the albedo (reflectivity) of solar radiation. In the 1970s when the prospect of global cooling was being studied, calculations were made of how much warming would be introduced by spreading carbon black on polar ice masses.

Table: Some physical characteristics of glacier ice on Earth.

	Antarctica (Grounded Ice)	Greenland	Glaciers and Small Ice Caps
Area (10^6 km^2)	11.97	1.68	0.55
Volume (10^6 km^3)	29.33	2.95	0.11
Mean Thickness (m)	2,488	1.575	200
Mean Elevation (m)	2,000	2,080	-
Equivalent Sea Level (m)	65	7	0.35
Accumulation (10^{12} kg/yr)	2200	535	-
Ablation (10^{12} kg/yr)	< 10	280	-
Calving (10^{12} kg/yr)	2200	255	-
Mean Equilibrium - Line Altitude (m)	-	950	0 - 6,300
Mass Turnover Time (yr)	~ 15,000	~ 5,000	50 - 1,000

1. Antarctica: The Antarctic ice sheet is a massive expanse of permanent ice whose high reflectance causes this region of the planet to lose more radiant energy through infrared radiation than it gains through illumination by the sun. This excess loss is made up by energy continually being supplied to the Antarctic region from lower latitudes by the atmosphere and ocean. Ice forms at high elevation on the Antarctic continent and slides downslope to the sea to create large floating icebergs that drift away from the polar ice mass. The ocean surface area covered by these icebergs reduces the amount of Open Ocean and reduces total evaporation and loss of heat energy from Polar Regions. This important fact only recently is being included in global climate models.

The Antarctic continent, because of its circularly symmetrical shape, high elevation, and extreme cold temperatures, leads to the formation of stationary and persistent stratospheric clouds that supply the conditions for ozone depletion. Like the global ocean, the Antarctic ice mass introduces thermal inertia into the climate system.

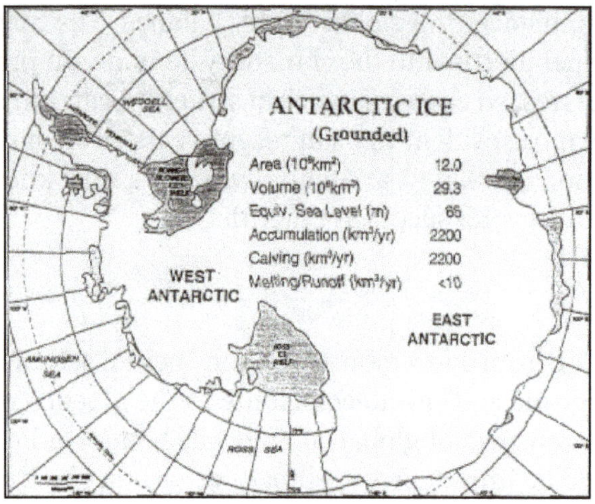

The Antarctic Sheet. (Gray areas are not covered by the ice sheet.)

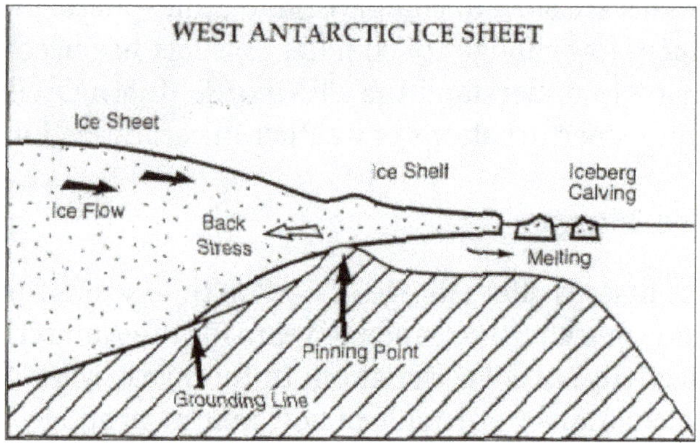

Processes in the West Antarctic ice sheet.

2. Greenland: The Greenland ice sheet is far less massive than the Antarctic ice field and is not nearly as cold. Having parts of its ice mass near the melting point, the Greenland ice field is more vulnerable to decrease in size than is the Antarctic ice sheet, and so its time scale for change is shorter than that of the Antarctic ice sheet.

Table: Annual mass balance of Greenland ice sheet in km³.

	Bader	Benson	Bauer	Weidick	Reeh
Accumulation	+ 630	+ 500	+ 500	+ 500 ± 100	+ 487
Melting	- 120 to - 270	-272	- 330	- 295 ± 100	- 169
Iceberg Discharge	- 240	-215	- 280	- 205 ± 60	- 318
Net Balance	+ 270 to + 120	+ 13	- 110	0	0

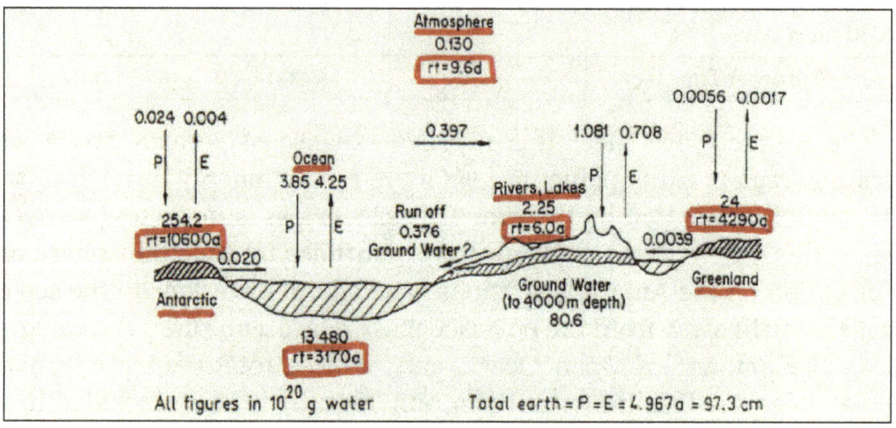

3. Mountain glaciers: Mountain glaciers contribute only a small amount of H2O to the global total, but their locations closer to population centers make them targets for urban and agricultural water supplies. They can experience noticeable change over periods of decades and, indeed, serve as independent indicators of global warming as can be seen in the accompanying plot of changes in glacier termini.

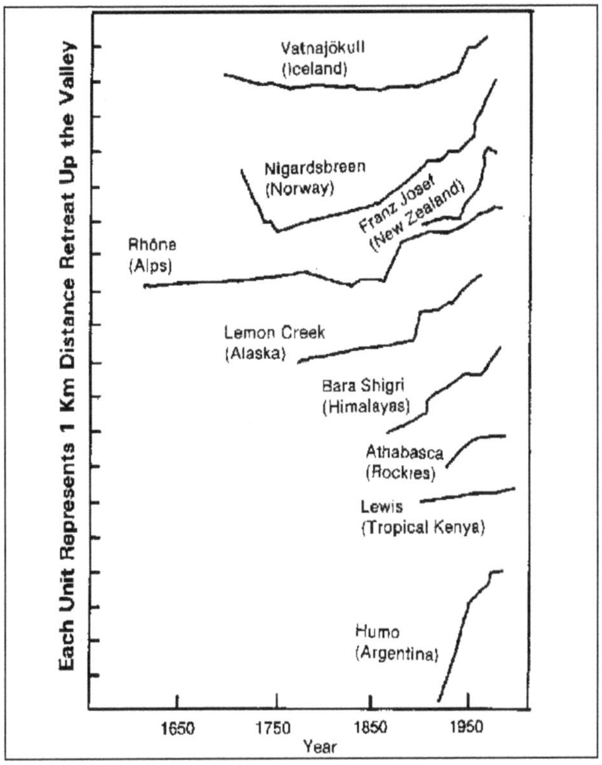

Retreat of Mountain Glaciers.

4. Arctic ice: Arctic ice is not anchored to land as is Antarctic ice. As floating ice, it governs the evaporation and heat loss from ocean water and having thickness far less than Antarctic or Greenland ice, it covers a relatively large surface area despite its relatively small volume. Being of lower elevation and being surrounded by asymmetric distributions of land and ocean, the North Pole does not provide an ideal location for persistent polar stratospheric clouds, although some ozone loss has been detected in this region.

Continental Seasonal Snow

Continental seasonal snow also provides recharge for reservoirs used for urban and agricultural purposes. These areas frequently also serve as important recreational sites.

Surface Fresh Water

The need of fresh water for humans and terrestrial animals, and the high cost of desalinization have put an increasing demand on global supplies of fresh water. In addition to its uses for direct consumption by humans and animals, and for cleaning, agriculture, and power production, fresh water bodies provide a source of food and ecosystem services previously described. River valleys running through dry regions form riparian zones that have much higher biological diversity than the arid surroundings. Rivers offer an inexpensive form of transportation: a single barge tow can carry the equivalent amount of grain or coal as 900 large trucks. The role of rivers and lakes for recreation and aesthetics should not be overlooked. Their contribution to a high "quality of life" is difficult to quantify but is increasingly recognized in societies having substantial leisure time.

Marshes and wetlands are bodies of fresh surface water that have particularly rich ecological diversity. These bodies support not only their own diversity but provide an ecological service in their ability to break down anthropogenic chemicals such as agricultural fertilizers and pesticides.

Subsurface Water

Soil moisture, permafrost, groundwater, and water in deep aquifers compose different forms of subsurface water that impact or are impacted by global change. Soil moisture provides the basic nutrient for plant growth and a reservoir for precipitation storage. By supplying moisture to the soil surface and to plants at a regulated rate depending on soil characteristics, soil moisture also influences surface evaporation and evapotranspiration (water loss by plants) and hence the global energy budget. By slowly delivering water to the surface and to plants over a period of weeks to months, soil moisture introduces a seasonal timescale into the climate system. A climate model with an inappropriate soil moisture submodel may allow its simulated soil to dry out too fast or too slow and cause, respectively, excessive or insufficient heating at the soil surface.

Permafrost plays an important role in global change because of the plant material it has locked up in ice. Melting permafrost exposes new plant material to temperatures that accelerate the decay process. This melting, from whatever cause, will lead to new releases of CO_2 and methane, both greenhouse gases, that lead to further global warming (a positive feedback to the climate system).

Ground water includes near surface aquifers that frequently are tapped by wells for human consumption, agricultural irrigation, and industrial uses. Their proximity to the surface and occasional direct connection to surface water, leave ground water supplies vulnerable to contamination from surface pollutants.

Deep aquifers represent water bodies that have been created thousands or more years ago by slow geo-climatic processes and their water is sometimes referred to as "fossil water". The large Ogallala aquifer in the Central Plains of the US is an example of a deep aquifer.

Volcanic eruptions cause episodic additions of rather small amounts of subsurface water to the atmosphere, but large eruptions put combinations of sulfates and water vapor into the stratosphere where they contribute to a transient (1-2 year) cooling effect on the global climate.

Biospheric Water

Biospheric water contributes little to total global water reservoirs but serves important functions of transporting nutrients in plants, keeping plants (and hence the earth's surface) cool and green (thereby influencing reflection of solar radiation), and serving as a source of nourishment for predators. Increased atmospheric CO_2 increases the water use efficiency of plants, which means they transpire less to the atmosphere. Less transpiring water means less evaporation and therefore warmer vegetation surfaces. Murray calls this the second greenhouse effect (the first being the direct warming due to greenhouse gases) and he estimates it could lead to an additional 2 °C to 5 °C warming of vegetation surfaces. Plants increase surface evaporation and sublimation (vaporization of snow) by suspending moisture in the form of dew, rainwater, or snow above the ground, thereby exposing much larger moisture surfaces to air movement and phase change.

Atmospheric Water

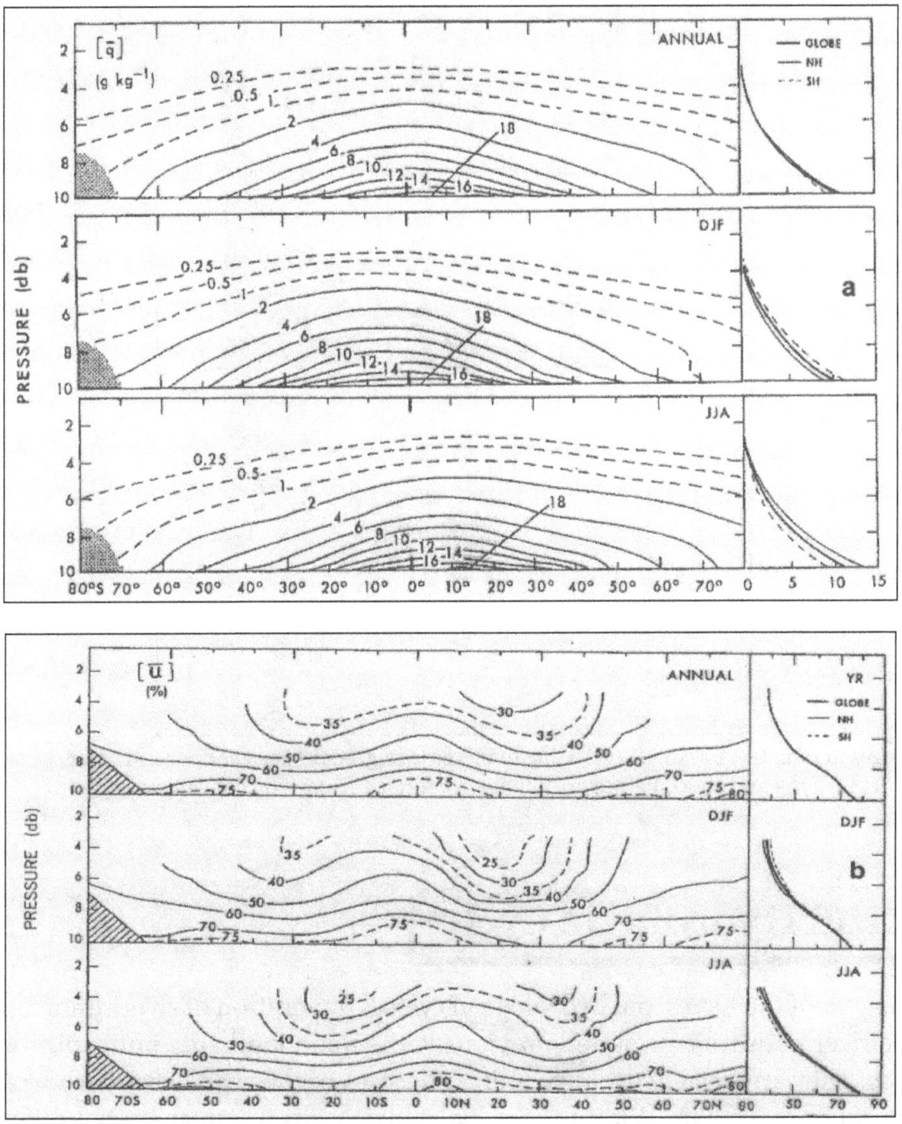

(a) Zonal-mean cross sections of the specific humidity in g kg^{-1} for annual, DJF, and JJA mean conditions. Vertical profiles of the hemispheric and global mean values are shown on the right. (b) Zonal-mean cross section of the relativity humidity in % for annual, DJF, and JJA mean conditions.

Atmospheric water consists of water vapor and liquid and ice components of clouds. In vapor form, water contributes to the greenhouse effect. The global distribution of water vapor is shown in previous figure in terms of g/kg of air, and in a second graph in terms of relative humidity.

Clouds

Clouds contribute both to global warming by absorbing infrared radiation from the surface and to global cooling by reflecting solar radiation. They also contribute to the pole ward transport of sensible heat. Atmospheric water is an important mechanism for cleansing the atmosphere of natural and anthropogenic contaminants. One such anthropogenic contaminant, sulfate aerosol, interacts with clouds to cause increased reflection of solar radiation and hence lead to global cooling.

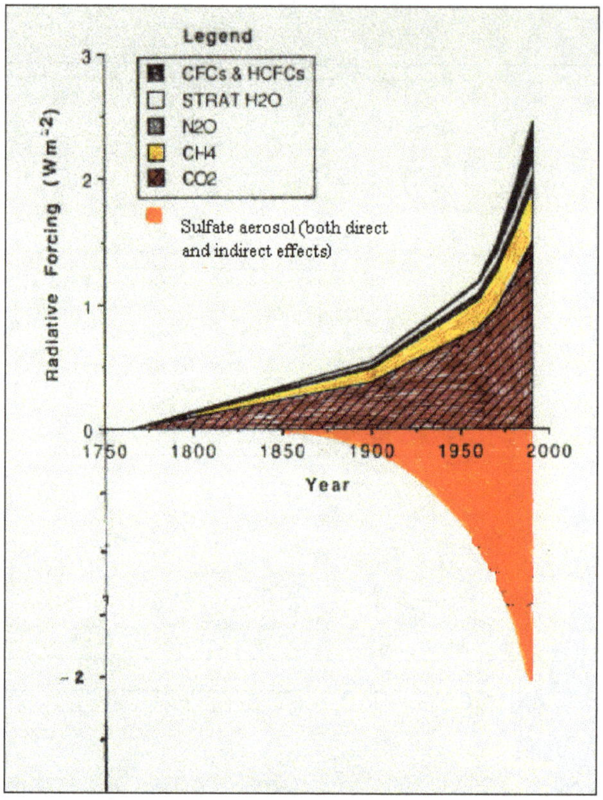

Changes in radiative forcing (Wm^{-2}) due to increase in greenhouse gas concentrations between 1765 and 1990. Values are changes in forcing from 1765 concentrations.

Terrestrial Hydrologic Cycle

The key component of the terrestrial hydrological cycle is generation of river runoff and movement of water in the river networks. As stated previously, the main land area units may be river basins and watersheds. They are ideal hydrologic units. The sizes of these areas vary from tens of sw.kms to several thousand square km. Within these hydrologic units, distinct spatial differences, in topography, climate, geology, structure, vegetation, soil properties, land use, land cover and other features may occur. The terrestrial hydrological cycle consider all aspects of the main processes, and factors that control the processes as well.

The main land area unit where this process occurs is the river watersheds. The sizes of these areas vary from tens of to 6900 square km (the Amazon River catchment area). Within these areas, distinct spatial differences, in topography, geology, vegetation, soil properties, land use, and meteorological conditions may be well-expressed even on small scales. The land surface heterogeneity may be essentially strengthened by human activities that can cause a significant modification of the characteristics of the natural landscapes. Therefore, to describe the terrestrial hydrological cycle it is important not only to single out the main processes, but also to take into account the relevant topographic, geological, vegetation, and soil parameters that control runoff generation conditions and give an opportunity to represent the land surface heterogeneity.

Components of Hydrologic Cycle

The circulation of water masses seen in all spheres of the earth involves several causative factors and components. The major components (or) elements of the hydrologic cycle are:

Precipitation

Precipitation is any liquid or frozen water that forms in the atmosphere and falls back to the Earth. It comes in many forms, like rain, sleet, and snow. Along with evaporation and condensation, precipitation is one of the three major parts of the global water cycle.

Precipitation forms in the clouds when water vapor condenses into bigger and bigger droplets of water. When the drops are heavy enough, they fall to the Earth. If a cloud is colder, like it would be at higher altitudes, the water droplets may freeze to form ice. These ice crystals then fall to the Earth as snow, hail, or rain, depending on the temperature within the cloud and at the Earth's surface. Most rain actually begins as snow high in the clouds. As the snowflakes fall through warmer air, they become raindrops.

Particles of dust or smoke in the atmosphere are essential for precipitation. These particles, called "condensation nuclei," provide a surface for water vapor to condense upon. This helps water droplets gather together and become large enough to fall to the Earth.

A common misconception is that when raindrops fall, they have a teardrop shape. In fact, smaller raindrops (ones that are approximately 1 millimeter (0.039 inches) across) are almost perfectly spherical. Larger raindrops (2–3 millimeters (.078-.118 inches) across) are also round, but with a small indent on their bottom side. They look more like kidney beans when falling to the Earth. Very large rain drops (larger than 4.5 millimeters (.177 inches)) have a huge indent and look more like a parachute. These extra-large drops usually end up splitting into two smaller droplets. The indents on raindrops are caused by air resistance.

Precipitation is always fresh water, even when the water originated from the ocean. This is because sea salt does not evaporate with water. However, in some cases, pollutants in the atmosphere can contaminate water droplets before they fall to the Earth. The precipitation that results from this is called acid rain. Acid rain does not harm humans directly, but it can make lakes and streams more acidic. This harms aquatic ecosystems because plants and animals often cannot adapt to the acidity.

Forms of Precipitation

Precipitation occurs in many forms e.g. drizzle, rain, glaze, sleet, snow, hail, dew and frost, depending upon the causes and temperature at the time of formation. Dew is condensation on the ground of atmospheric vapor caused by radiational cooling of the lower layers of atmosphere, usually at night. Frost is dew formed under freezing conditions. Dew and frost are quantitatively unimportant and rarely measured.

- Drizzle: Drop size < 0.5 mm in dia. and intensity is usually < 1 mm/hr and generally occurs in conjunction with warm frontal lifting.

- Rain: Drop size is between 0.5 to 6 mm in dia. Drops bigger than 6 mm tend to break up as they fell. It is formed by condensation and coalescence of cloud droplets at temperatures above the freezing point.
- Glaze: It is the ice coating formed when drizzle or rain freezes as it comes in contact with cold objects on the ground.
- Sleet: It is frozen raindrops cooled to ice stage while falling through air at sub-freezing temperature.
- Snow: It is a precipitation in the form of ice crystals resulting from sublimation, i.e., directly from water vapor to ice.
- Snow Flake: It is made of a number of ice crystals fused to gather.
- Hail: It is precipitation in the form of balls or lumps of ice over 5 mm diameter formed by alternate freezing and melting as they are carried up and down in highly turbulent air currents.

Mechanisms for Production of Rainfall

The following four conditions are necessary for the production of rainfall:

- Mechanism to produce cooling of the air: The pressure reduction due to ascending air from surface to upper levels in the atmosphere is the only known mechanism capable of producing large drops in the temperature.
- Mechanism to produce condensation: Condensation in the atmosphere takes place on "hygroscopic nuclei" small particles of substances that have an affinity for water.
- Mechanism for droplet growth: A tendency for the droplets to remain small and therefore to resist falling is called "colloidal stability". The most effective processes for droplet enlargement are:
    - The difference in speeds between large and small droplets.
    - The co-existence of ice crystals and water droplets.
- To produce accumulation of moisture of sufficient intensity to account for the observed rates of rainfall: Regardless of whether or not the other conditions for precipitation are fulfilled, continuity considerations demand that there must be a good amount of moisture present in the atmosphere so that evaporation losses between ground and cloud be compensated, if there is to be appreciable rain.

Types of Precipitation

There are three major types of precipitation: cyclonic, convective, and orographic. Each type represents a different method of lifting an air mass, resulting in cooling and condensation of atmospheric water vapor.

Cyclonic Precipitation

It is caused by lifting associated with the horizontal convergence of inflowing atmosphere into an area of low pressure. There are two kinds of cyclonic precipitation. Non-frontal precipitation involves only this convergence and lifting. Frontal precipitation results when one air mass is lifted over another. A front is defined as the boundary between two air masses of different temperatures and densities.

A warm front is the result of a warm air mass overriding a cold air mass, causing extensive areas of cloudiness and precipitation. As the warm front approaches a given area, the precipitation becomes more continuous and intense. Warm fronts move at a speed of 15-50 km/h (10-30 mph).

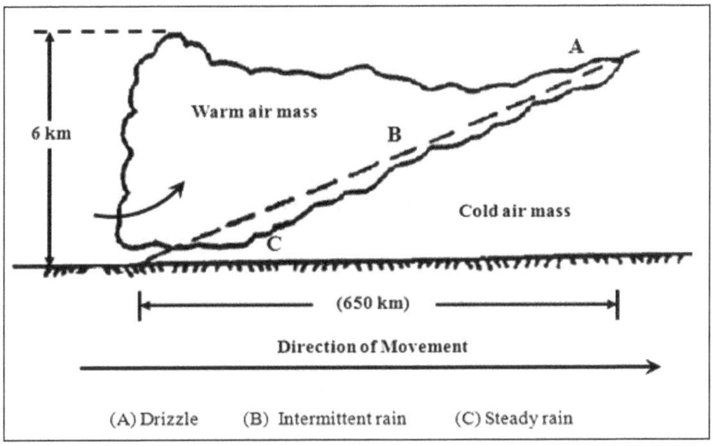

Warm Front

A cold front results from a strong push of a cold air mass against and beneath a warm air mass. At the front towering clouds develop together with intense short duration precipitation. Cold fronts move at a speed of 30-80 km/h (20-50 mph).

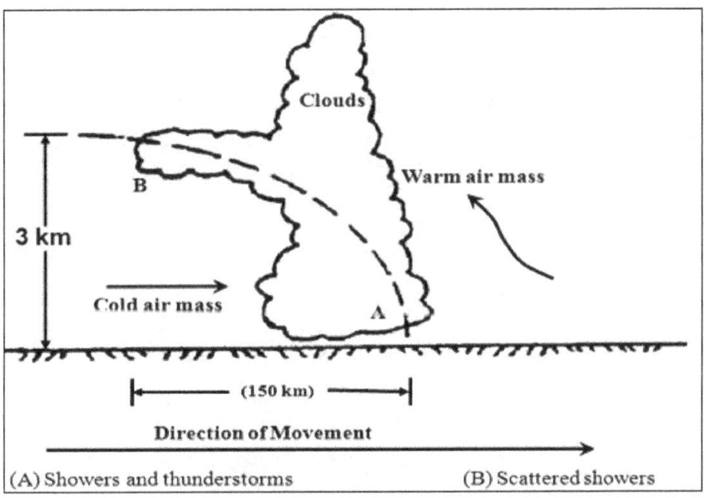

Cold Front

An occluded front occurs when a cold front overtakes a warm front. The precipitation pattern is a combination of both warm and cold frontal distribution. Occluded fronts move at a speed of from 8-50 km/h (5-30 mph).

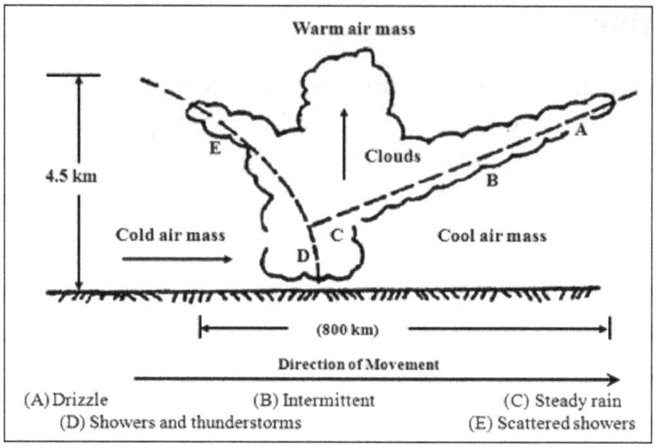

Occluded Front

Convective Precipitation

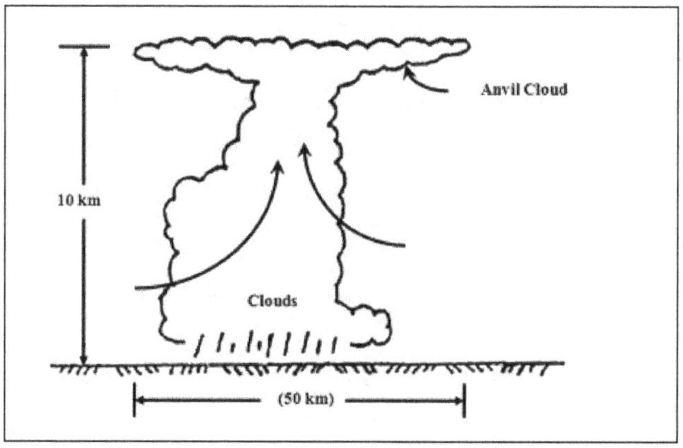

Thunderstorm

It results when air that is warmer than its surrounding rises and cools. The precipitation is of a shower type, varying from light showers to cloudbursts. The typical thunderstorms resulting from heating of the atmosphere in the afternoon hours is the best example of convective rainfall. Thunderstorms occur throughout the world, especially in the summer. They are the characteristic form of rain in the tropics, wherever cyclonic circulation does not operate. A cross section through a typical thunderstorm is shown in the figure.

Orographic Precipitation

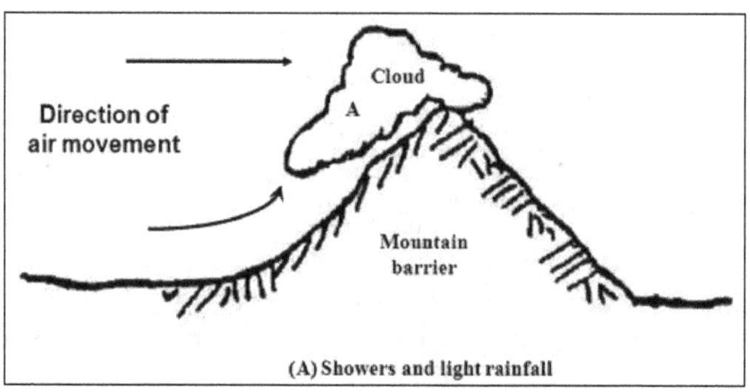

It is caused when air masses are lifted as they move over mountain barriers. Such orographic barriers tend to increase both cyclonic and orographic precipitation due to the increased lifting involved. Precipitation is generally heavier on the windward slope than on the leeward slope.

Evaporation

Evaporation is the process by which water changes from a liquid to a gas or vapor. Evaporation is the primary pathway that water moves from the liquid state back into the water cycle as atmospheric water vapor. Studies have shown that the oceans, seas, lakes, and rivers provide nearly 90 percent of the moisture in the atmosphere via evaporation, with the remaining 10 percent being contributed by plant transpiration.

A very small amount of water vapor enters the atmosphere through sublimation, the process by which water changes from a solid (ice or snow) to a gas, bypassing the liquid phase. This often happens in the Rocky Mountains as dry and warm Chinook winds blow in from the Pacific in late winter and early spring. When a Chinook takes effect local temperatures rise dramatically in a matter of hours. When the dry air hits the snow, it changes the snow directly into water vapor, bypassing the liquid phase. Sublimation is a common way for snow to disappear quickly in arid climates.

Cooling towers at a power-production facility. Power-generation plants, including Plant Bowen in Georgia, produce power by using heat (in this case, from burning coal) to convert water into steam.

One very significant by-product of thermoelectric facilities is heat from the power-production equipment. Plants withdraw lots of waste, use it to cool the equipment, and then need to release used water back into the environment. Releasing hot water back into rivers would harm the ecology, so many power plants have tremendous cooling towers, where hot water is sprayed inside and evaporation is used to cool the release water before it goes back into the environment.

Why Evaporation Occurs

Heat (energy) is necessary for evaporation to occur. Energy is used to break the bonds that hold water molecules together, which is why water easily evaporates at the boiling point (212° F, 100° C) but evaporates much more slowly at the freezing point. Net evaporation occurs when the rate of evaporation exceeds the rate of condensation. A state of saturation exists when these two process rates are equal, at which point the relative humidity of the air is 100 percent. Condensation, the opposite of evaporation, occurs when saturated air is cooled below the dew point (the temperature to which air must be cooled at a constant pressure for it to become fully saturated with water), such as on the outside of a glass of ice water. In fact, the process of evaporation removes heat from the environment, which is why water evaporating from your skin cools you.

Evaporation Drives the Water Cycle

Evaporation from the oceans is the primary mechanism supporting the surface-to-atmosphere portion of the water cycle. After all, the large surface area of the oceans (over 70 percent of the Earth's surface is covered by the oceans) provides the opportunity for large-scale evaporation to occur. On a global scale, the amount of water evaporating is about the same as the amount of water delivered to the Earth as precipitation. This does vary geographically, though. Evaporation is more prevalent over the oceans than precipitation, while over the land, precipitation routinely exceeds evaporation. Most of the water that evaporates from the oceans falls back into the oceans as precipitation. Only about 10 percent of the water evaporated from the oceans is transported over land and falls as precipitation. Once evaporated, a water molecule spends about 10 days in the air. The process of evaporation is so great that without precipitation runoff, and groundwater discharge from aquifers, oceans would become nearly empty.

People Make use of Evaporation

If you ever find yourself stranded on an island in need of some salt, just grab a bowl, add some seawater, and wait for the sun to evaporate the water. In fact, one way to produce table salt is to evaporate saline water in evaporation ponds, a technique used by people for thousands of years.

Your table salt might have come from an evaporation pond.

Seawater contains other valuable minerals that are easily obtained by evaporation. The Dead Sea is located in the Middle East within a closed watershed and without any means of outflow, which is abnormal for most lakes. The primary mechanism for water to leave the lake is by evaporation, which can be quite high in a desert—upwards of 1,300 - 1,600 millimeters per year. The result is that the waters of the Dead Sea have the highest salinity and density (which is why you float "higher"; when you lay in saline water) of any sea in the world, too high to support life. The water is ideal for locating evaporation ponds for the extraction of not only table salt, but also magnesium, potash, and bromine.

Evaporation Process

Evaporation includes all processes by which water returns to the atmosphere as water vapour: evaporation of intercepted rain and snow; evaporation from bare soil and water bodies, such as ponds, lakes, and streams; and transpiration from plant leaves. Evaporation (and Transpiration) are small for a runoff event and can be neglected. The bulk of these abstractions take place during the time between runoff events, which is usually long. Hence, these are more important during this time interval. Evaporation requires the following four conditions:

- Available water.
- Higher humidity at the evaporative surface (i.e., vapour pressure) than in the surrounding air.
- Energy to evaporate the water.
- Movement, or transfer, of water vapour away from the evaporative surface.

Energy required to evaporate water depends on incoming solar radiation, reflectivity of the evaporative surface, and air and surface temperature. Diffusion and convection move the vapour away from the surface. Increasing solar radiation, air temperature, and wind speed and decreasing atmospheric humidity all create an increase in evaporation rate. Evaporation is enhanced by warm air flowing over a cooler surface (e.g., air moving from dry rangeland over an irrigated crop or a small lake), but decreases rapidly with distance from the boundary between dry and wet surfaces. Intercepted rain or snow, and open water are in direct contact with the air. Both boundary-layer and aerodynamic resistance affect water loss from these surfaces. The boundary layer is a thin layer adjacent to a surface through which vapour moves by diffusion. Aerodynamic resistance describes vapour movement in the rest of the atmosphere. Both resistances depend on the size and shape of the evaporative surface, and both decrease as wind speed increases. Tree needles have a lower boundary layer resistance than large leaves and a much lower resistance than that of a lake. Trees generate more turbulence to airflow than smooth surfaces, such as a lake; consequently, trees have lower aerodynamic resistances at the same wind speed. The combined resistances for a wet surface are relatively low compared to the resistance to movement of water from inside leaves or from below a dry soil surface.

Factors Affecting Evaporation

The factors that affect evaporation are:

- Wind: When wind speed is high it assists evaporation.
- Heat: Evaporation is more in summer as compared to winter.
- Exposed surface area: For instance, a wet cloth spread out dries faster than when folded.
- Humidity: Dryness assists evaporation; for instance, clothes dry faster in summer than during the monsoon when the air is humid.

- Nature of the liquid: Rate of evaporation depends upon the type of liquid; for example, petrol evaporates faster than water.
- Vapour pressure: If pressure is applied on the surface of a liquid, evaporation is hindered; consider, for example, the case of a pressure cooker.

Measurement of Evaporation

Lysimeter

A lysimeter is a measuring device which can be used to measure the amount of actual evapotranspiration which is released by plants, usually crops or trees. By recording the amount of precipitation that an area receives and the amount lost through the soil, the amount of water lost to evapotranspiration can be calculated.In general, a lysimeter consists of the soil-filled inner container and retaining walls or an outer container, as well as special devices for measuring percolation and changes in the soil-moisture content. There is no universal international standard lysimeter for measuring evapotranspiration. The surface area of lysimeters in use varies from 0.05 to some 100 m2 and their depth varies from 0.1 to 5 m.

According to their method of operation, lysimeters can be classified into non-weighable and weighable instruments. Each of these devices has its special merits and drawbacks, and the choice of any type of lysimeter depends on the problem to be studied. Monolithic weighable lysimeters are a tool for water balance studies and solute transport determination. Lysimeters are of two types:

- Non-weighable (percolation-type) lysimeters can be used only for long-term measurements, unless the soil-moisture content can be measured by some independent and reliable technique. Large-area percolation-type lysimeters are used for water budget and evapotranspiration studies of tall, deep rooting vegetation cover, such as mature trees. Small, simple types of lysimeters in areas with bare soil or grass and crop cover could provide useful results for practical purposes under humid conditions. This type of lysimeter can easily be installed and maintained at a low cost and is, therefore, suitable for network operations.
- Weighable lysimeters, unless of a simple microlysimeter-type for soil evaporation, are much more expensive, but their advantage is that they secure reliable and precise estimates of short-term values of evapotranspiration, provided that the necessary design, operation and siting precautions have been taken.

Several weighing techniques using mechanical or hydraulic principles have been developed. The simpler, small lysimeters are usually lifted out of their sockets and transferred to mechanical scales by means of mobile cranes. The container of a lysimeter can be mounted on a permanently installed mechanical scale for continuous recording. The design of the weighing and recording system can be considerably simplified by using load cells with strain gauges of variable electrical resistance. The hydraulic weighing systems use the principle of fluid displacement resulting from the changing buoyancy of a floating container (so called floating lysimeter), or the principle of fluid pressure changes in hydraulic load cells. The large weighable and recording lysimeters are recommended for precision measurements in research centres and for standardization and parameterization of other methods of evapotranspiration measurement and the modelling of evapotranspiration. Small weighable types of lysimeters are quite useful and suitable for network operation. Microlysimeters for soil evaporation are a relatively new phenomenon.

Pan Evaporation

Pan evaporation is a measurement that combines or integrates the effects of several climate elements: temperature, humidity, rain fall, drought dispersion, solar radiation, and wind. Evaporation is greatest on hot, windy, dry, sunny days; and is greatly reduced when clouds block the sun and when air is cool, calm, and humid. Pan evaporation measurements enable farmers and ranchers to understand how much water their crops will need. There are many types of evaporation pans used by farmers. However, the universal pan is the United States Weather Bureau (USWB) Class A pan evaporimeter. It is important to use the same dimensions as this universal pan, mainly because the effect of wind and temperature on evaporation will vary with the surface area and the depth of water in the pan. Evaporation and irrigation replacements cannot be compared between sites if non-standard pans are used.

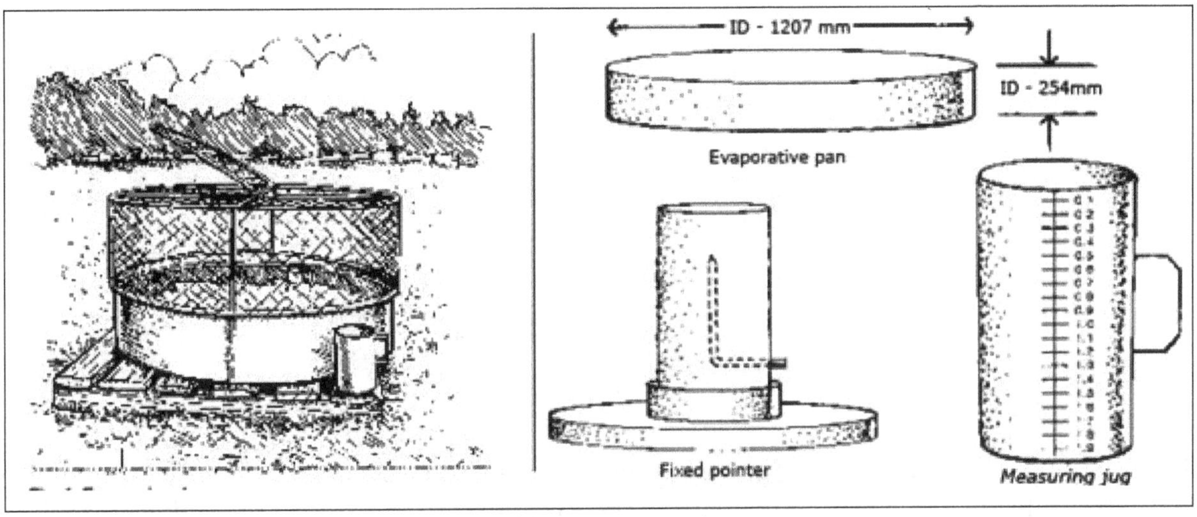

USWB Class A Pan Evaporimeter.

Construction

There are three parts to an evaporimeter. All parts can be made very cheaply with common materials. Alternatively a complete unit can be purchased at considerably greater cost. The following is a description of how to construct the three components of the evaporimeter.

Evaporation Pan: The evaporation pan must be made to the standard specifications of an internal diameter of 1207 mm and height of 254 mm using 20 gauge galvanised iron. The standard material is galvanised iron as alternatives will have different thermal and reflectance properties, therefore altering the evaporation rate. It is best to have the pan made by either a galvanised tank manufacturer or an engineering firm. Before the pan is sited in the field it should be checked for leaks.

Fixed Pointer

The fixed pointer that sits inside the pan can be made from standard irrigation fittings and a piece of stainless steel rod. There are three parts to the fixed pointer:

- The base, a 100 mm PVC flange.

- The pointer support, a 230 mm long piece of 100 mm PVC pipe. Four equally spaced 9 mm holes 70 mm from the base are drilled to allow the water height around the fixed pointer to quickly adjust to the water height in the pan. A single 15 mm long 5 mm wide elongated hole is also drilled 70 mm from the base of the PVC pipe.
- The pointer, a 170 mm long piece of 5 mm stainless steel rod bent at a right angle 60 mm from one end. From the shorter end a thread is tapped for about 15 mm and a point is ground on the other end of the rod.

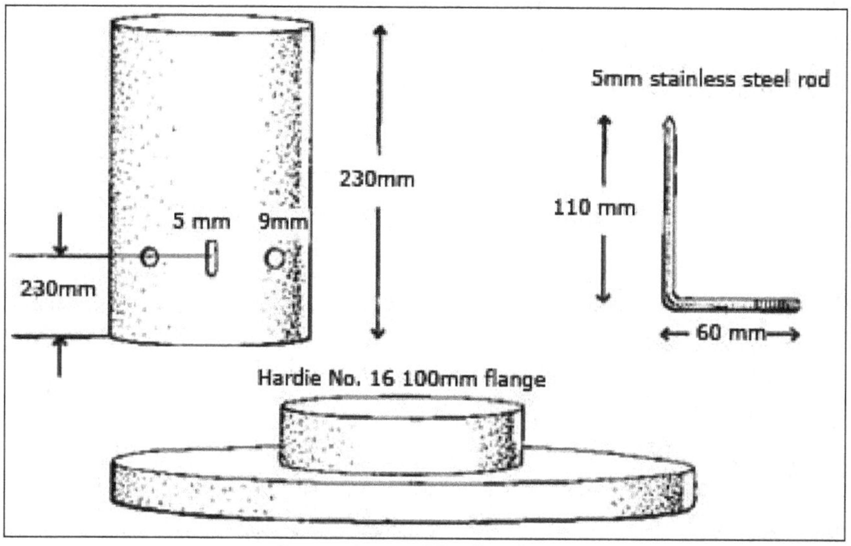

Components of fixed pointer.

After fitting the PVC pipe into the flange, the stainless steel rod is inserted into the elongated hole with nuts located on the inside and outside of the PVC pipe. To initially set the stainless steel rod in the correct position, the fixed pointer is placed in the pan and the pan is filled with water to a depth of 190 mm. The rod is then slid up or down in the 5 mm elongated hole so that the point of the rod just breaks the surface of the water.

Measuring Cylinder

To measure evaporation the pan must be refilled with a known volume of water. The surface area of the pan is 1.14 square metres, so for every mm of evaporation 1.14 litres of water must be added to the pan. A transparent plastic 2 litre measuring jug with vertical sides is an excellent measuring cylinder if it is scaled properly. It is important that the jug actually holds more water than 2 litres so the sides of the jug must extend past the 2 litre mark. The jug is filled with 2.28 litres of water and the water level marked. This can conveniently be done by weighing the jug and adding 2.28 kilograms of water. For most jugs this will just about overflow, which is perfect.

A jug of water filled to the marker will be equivalent to 2 mm of evaporation. To scale the jug when less than 2 mm of water is required to fill the pan, the distance from the top marker to the bottom of the jug is measured and divided by 20. The numbers 0 to 2.0 in increments of 0.1 are then written with a permanent marking pen from the top marker to the bottom of the jug. These numbers are equivalent to the same number of mm of evaporation from the pan.

Measurement

With evaporation the water level in the pan will fall. To measure the amount of evaporation, water is added to the pan with the measuring jug filled to the top mark. Water is added until the pointer just breaks the surface of the water. The PVC pipe supporting the pointer will help by reducing wave motion. It is important to keep track of the number of jugs used to refill the pan and the reading on the last jug when the pan water level is just broken by the pointer. The total amount of water added equals the amount of evaporation.

It is also essential to measure rainfall in conjunction with evaporation. Both measurements enable evaporation to be calculated on rainy days. After heavy rain the pan may have to be emptied to bring the water level down to the pointer. After rainfall on a hot summer day, less water may have to be removed than actually fell as rain. For example, after a 25 mm rainfall there might only be 12 mm of water removed from the pan with the measuring jug to bring the water level back to the pointer. The difference between the rainfall (25 mm) and the water removed from the pan (12 mm) is the evaporation. In this example it is 13 mm.

If the rain does not fill the pan above the pointer, the rainfall must still be added onto the measured evaporation to give the actual evaporation. For example, if there was 7 mm of rainfall and 6 mm of water was added to the pan with the measuring jug then the evaporation would be 13 mm.

Evaporation measurements should be routinely done every day at 9.00 am and clearly recorded. If measurements are not done routinely then the volume of water in the pan will decrease and take less time to heat up during the day and cool at night. This will induce an error which will become greater as the volume of water in the pan decreases. Evaporation measurements are very simple and take less than 5 minutes.

Determination of Evaporation from Water Surfaces

Evaporation from water surfaces can be determined by:

- Water budget,
- Energy budget,
- Mass transfer methods,
- Combination methods,
- Evaporation formulas.

Water Budget

The water-budget equation for estimating evaporation can be written as:

$$E = I + P - O - O_S + DS$$

Where,

- E =Evaporation

- I = Inflow
- P = Precipitation
- O = Outflow
- O_s = Seepage
- ΔS = Change in storage

Here, Inflow, outflow, precipitation, and change in storage can be measured reasonably accurately. Seepage, O_s, cannot be measured or evaluated directly and accurately, and the extent to which this quantity is accurate will affect the true value of evaporation.

The water-budget method of determining long-term evaporation can be used as a standard for comparing other methods. This method is not perfect, but it is satisfactory for practical purposes.

Transpiration

Transpiration is the process of releasing the water absorbed by the plants through their root system after utilizing the nutrients for building their tissues, in a specified time. Vegetation including numerous growing plants play a significant role in the hydrologic cycle. The water which is drawn into the plants rootlets from the soil moisture, owing to osmotic pressure moves up through the plants stems and leaves. Through the stomatal openings, the water is released out as water vapour. The amount of transpiration depends on the density and size of the vegetation existing in place. The amount of water used for irrigating the crops gets transpired into the air. Transpiration is dominant during the growing season of crops in agricultural lands. Most of this happens during day time, when photosynthesis is active in plants. Transpiration is limited due to the shortage of soil moisture is some places.

Variables Affecting Transpiration Rates

The rate at which transpiration occurs is varied based on temperature, air movement such as wind, how much moisture is in the soil and surrounding air, the type of plant and land use.

- Temperature plays a major role in the rate of transpiration. As the temperature increases, transpiration will increase due to a higher concentration in sunlight and warm air. However, if temperatures remain high for long periods of time eventually leading to drought, transpiration may go down to conserve water in the plant. Colder temperatures usually lead to very little or no transpiration occur whatsoever.
- Air movement such as wind can aid with transpiration. When the air is still or there is no wind, humidity may build up around the plant from transpiration eventually decreasing the amount of water being released. When there is wind present, this causes the air to be replaced constantly allowing the plant to transpire.
- Increased humidity has a reduces transpiration. The lower the humidity the easier it is for the plant to release water.

- The type of plant is also a factor in how much a plant will transpire. For example plants grow in arid hot areas such as the desert will transpire less as they aim to conserve water.
- Groundwater availability is the biggest factor for how much a plant transpires. During growing season, when higher volumes of water are entering the top layer of soil, it is much easier for a plant to absorb more water. If a plant is situated near a large body of water or for example on a slope where it may be closer to the water table, it will also be easier for the plant to acquire water. During the dormant or dry season, less water is readily available to plants thus drying out the plants. Figure shows how water table fluctuations effect transpiration.

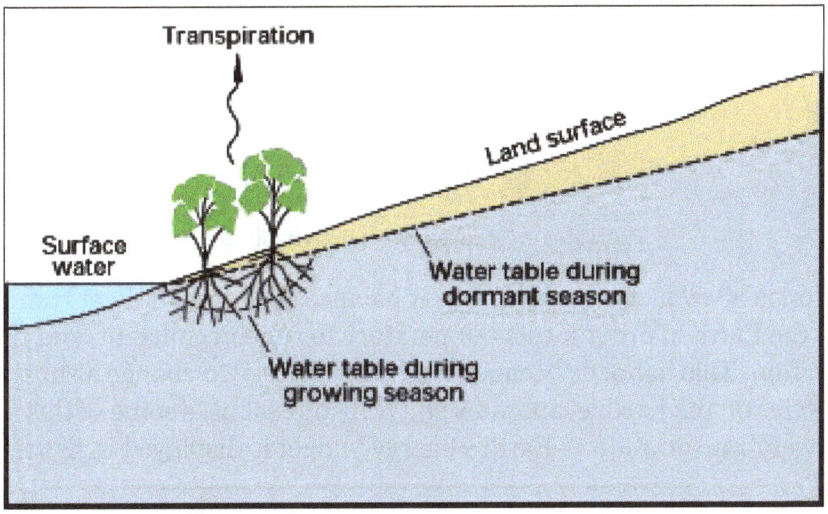

In areas where the water table is closer to the surface such as near large bodies of water or on angled terrain, plant roots can access the area beneath the water table; this creates an easier environment for plants to acquire water to aid with transpiration.

- The conversion of land by humans can have a detrimental effect on transpiration. When natural land is converted to agricultural land, the vegetation cover is diminished. This decreases transpiration resulting in erosion and an increase in rain runoff thus creating silt buildup in mass bodies of water.

Evapotranspiration

Evapotranspiration is the transfer of energy from the Earth's surface to the atmosphere in the form of latent heat, due to the evaporation of water from the ground and bodies of water, and the transpiration of water from plants. The transpiration aspect of evapotranspiration accounts for about 10% of the moisture in the atmosphere, with evaporation from oceans and other bodies of water accounting for nearly 90% and a tiny amount coming from sublimation (ice changing into water vapour without first becoming liquid).

Transpiration is essentially evaporation of the water from plant leaves. Plants must transpire just like humans have to breathe, and this process depends on many environmental factors. Evapotranspiration is an important part in the Earth's hydrologic cycle. About 30% of the total ocean volume evaporates each year. Wind carries this water vapour to higher elevations where it cools and condenses, and finally precipitates. About 76% of it goes right back to the ocean, and the rest hits land and drives the life cycles there.

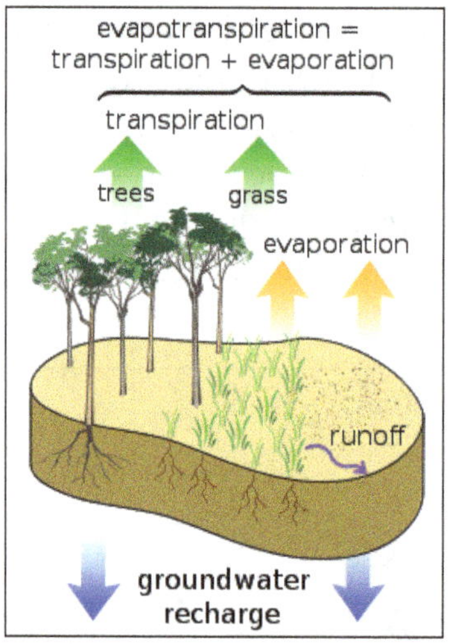

The water cycle relies on evapotranspiration.

Evapotranspiration is a key factor in Earth's heat balance. The radiant heat coming from the Sun must be rid of by the Earth in order for its temperature to remain constant, and this is partly done by evapotranspiration. This happens because it requires energy to change a substance from liquid to gas, so as the water on the Earth evaporates and transpires it uses some of this solar energy. The contribution of evapotranspiration to Earth's energy budget is displayed in figure.

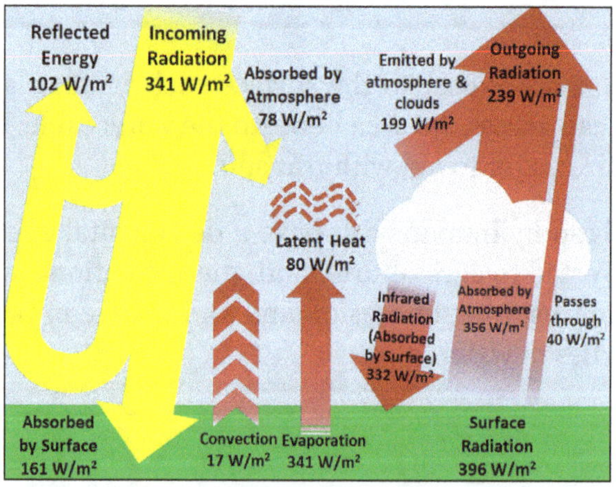

Detailed diagram showing incoming and outgoing energy in a balanced example of Earth's energy budget, with incoming and outgoing values of 341 W/m^2. Evapotranspiration accounts for 80 W/m^2 of the outgoing energy.

Estimating Evapotranspiration

Evapotranspiration can be measured or estimated using several methods:

Indirect Methods

Pan evaporation data can be used to estimate lake evaporation, but transpiration and evaporation of intercepted rain on vegetation are unknown. There are three general approaches to estimate evapotranspiration indirectly.

Catchment Water Balance

Evapotranspiration may be estimated by creating an equation of the water balance of a drainage basin. The equation balances the change in water stored within the basin (S) with inputs and outgoes:

$$\Delta S = P - ET - Q - D$$

The input is precipitation (P) and the outputs are evapotranspiration (which is to be estimated), streamflow (Q), and groundwater recharge (D). If the change in storage, precipitation, streamflow, and groundwater recharge are all estimated, the missing flux, ET, can be estimated by rearranging the above equation as follows:

$$ET = P - \Delta S - Q - D$$

Energy Balance

A third methodology to estimate the actual evapotranspiration is the use of the energy balance.

$$\lambda E = R_n - G - H$$

Where, λE is the energy needed to change the phase of water from liquid to gas, R_n is the net radiation, G is the soil heat flux and H is the sensible heat flux. Using instruments like a scintillometer, soil heat flux plates or radiation meters, the components of the energy balance can be calculated and the energy available for actual evapotranspiration can be solved.

The SEBAL and METRIC algorithms solve the energy balance at the earth's surface using satellite imagery. This allows for both actual and potential evapotranspiration to be calculated on a pixel-by-pixel basis. Evapotranspiration is a key indicator for water management and irrigation performance. SEBAL and METRIC can map these key indicators in time and space, for days, weeks or years.

Experimental Methods for Measuring Evapotranspiration

One method for measuring evapotranspiration is with a weighing lysimeter. The weight of a soil column is measured continuously and the change in storage of water in the soil is modeled by the change in weight. The change in weight is converted to units of length using the surface area of the weighing lysimeter and the unit weight of water. Evapotranspiration is computed as the change in weight plus rainfall minus percolation.

Eddy Covariance

The most direct method of measuring evapotranspiration is with the eddy covariance technique in which fast fluctuations of vertical wind speed are correlated with fast fluctuations in atmospheric water vapor density. This directly estimates the transfer of water vapor (evapotranspiration) from the land (or canopy) surface to the atmosphere.

Hydrometeorological Equations

The most general and widely used equation for calculating reference ET is the Penman equation. The Penman-Monteith variation is recommended by the Food and Agriculture Organization and the American Society of Civil Engineers. The simpler Blaney-Criddle equation was popular in the Western United States for many years but it is not as accurate in regions with higher humidities. Other solution used includes Makkink, which is simple but must be calibrated to a specific location, and Hargreaves.

To convert the reference evapotranspiration to actual crop evapotranspiration, a crop coefficient and a stress coefficient must be used. Crop coefficients as used in many hydrological models usually change along the year to accommodate to the fact that crops are seasonal and, in general, plants behave differently along the seasons: perennial plants mature over multiple seasons, and stress responses can significantly depend upon many aspects of plant condition.

Potential Evapotranspiration

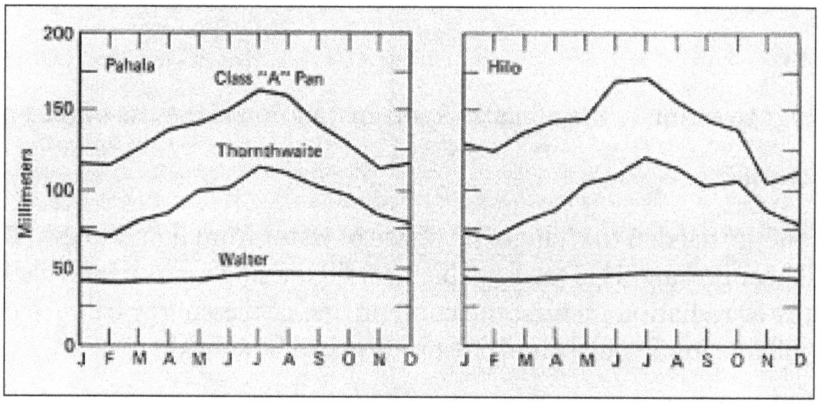

Monthly estimated potential evapotranspiration and measured pan evaporation for two locations in Hawaii, Hilo and Pahala.

Potential evapotranspiration (PET) is the amount of water that would be evaporated and transpired by a specific crop or ecosystem if there were sufficient water available. This demand incorporates the energy available for evaporation and the ability of the lower atmosphere to transport evaporated moisture away from the land surface. Potential evapotranspiration is higher in the summer, on less cloudy days, and closer to the equator, because of the higher levels of solar radiation that provides the energy for evaporation. Potential evapotranspiration is also higher on windy days because the evaporated moisture can be quickly moved from the ground or plant surface, allowing more evaporation to fill its place.

Potential evapotranspiration is expressed in terms of a depth of water, and can be graphed during the year. Potential evapotranspiration is usually measured indirectly, from other climatic factors, but also depends on the surface type, such as free water (for lakes and oceans), the soil type for bare soil, and the vegetation. Often a value for the potential evapotranspiration is calculated at a nearby climate station on a reference surface, conventionally short grass. This value is called the reference evapotranspiration, and can be converted to a potential evapotranspiration by multiplying with a surface coefficient. In agriculture, this is called a crop coefficient. The difference between potential evapotranspiration and precipitation is used in irrigation scheduling.

Average annual potential evapotranspiration is often compared to average annual precipitation, P. The ratio of the two, P/PET, is the aridity index. A humid subtropical climate is a zone of climate characterized by hot and humid summers, and cold to mild winters Subarctic regions fall between 50 °N and 70 °N latitude, depending on local climates. Precipitation is low, and vegetation is characteristic of the coniferous/taiga forest.

Factors Affecting Evapotranspiration (ET)

Weather parameters, crop characteristics, management and environmental aspects are factors affecting E, T, and ET. Details of these factors are as follows:

Climatic Factors that Affecting ET

Radiation

Radiation is the main source of energy for the ET process. It depends on the global solar radiation flux density and vegetation albedo. Darker vegetation absorbs more incident solar radiation and evapotranspires more. A thorough understanding of the factors controlling the energy balance of a cropped soil enables making accurate estimates or predictions of ET and irrigation water requirements. It also facilitates more effective irrigation water management.

Temperature

Over the course of a day, an increase of the air temperature causes an increase on the saturation deficit triggering a higher evaporative demand in the air, and leading to high ET rates.

Relative Humidity (%RH)

Air relative humidity acts in conjunction with temperature. The higher relative humidity, the lesser the evaporative demand and, therefore, the lower ET.

Wind

Advection represents the horizontal transport of energy from a drier area to another more humid, and such additional energy is utilized in the ET process. Wind also helps remove water vapor near the plants to other regions.

Crop Factors that Affecting ET

Crop factors such as crop species, radiation reflection coefficient, leaf area index (LAI) in different growth stages of the plant, plant height and root depth (depth of the radicular system) are influenced on the crop ET, which are described details as follows:

Crop Species

This factor is related to the foliar architecture (spatial distribution of the leaves), internal resistance of the plant to water transport, and other morphological aspects (number, size, and distribution of stomata, etc.), which exert a direct influence on ET.

Radiation Reflection Coefficient

Radiation reflection influences directly net radiation availability for the ET process. The darker the vegetation, the lower the reflection coefficient and the higher net radiation.

Leaf Area Index (LAI) in Different Growth Stages of Plant

LAI in different growth stages of a plant is directly related to the size of the transpiring foliar surface, for the larger leaf area the larger the transpiring surface, and the higher the potential for water use.

Plant Height

Plant height also influences the ET. Taller and rougher plants interact more efficiently with the atmosphere in motion, extracting more energy from the air and, therefore, increasing ET.

Rooting Depth of Plant (Depth of the Radicular System)

Rooting depth of plant is directly related to the volume of soil explored by the roots, aiming at meeting the atmospheric hydric demand. A superficial radicular system, for exploring a smaller soil volume, keeps the crop more susceptible to drying periods.

Crop Management and Growing Environmental Conditions also Influence the ET

The following crop management and growing environmental conditions are influenced the crop ET:

Row to Row or Plant to Plant Spacing

Usually, intraspecific competition is found between plants/crops of the same species/types for their essential growth elements. A limited spacing between the plant to plant or row to row of the same species or different species of plants/crops, consequences in an intense competition for water, light, nutrient, etc., causing as a consequence an increase on ET.

Crop Orientation

Crops oriented perpendicularly to predominant winds tend to extract more energy from the air than those oriented in parallel. For regions with constant winds, a solution to prevent the stomata-closing would be the use of windbreaks. A windbreak reduces wind velocities and decreases the ET rate of the field directly beyond the barrier.

Soil Properties (Structure and Texture)

Soil texture and soil structure are both unique properties of the soil that will have a profound effect on the behavior of soils. Both the properties influence the crop ET through influencing the water holding capacity of the soils. Clay soils have higher water holding capacity than sandy soils and are proficient of preserving a more persistent crop ET rate for longer.

Chemical/Physical Impediments

Inhibitions system limit the growth of the radicular system of a plant, affecting the root system of plants to explore a wider volume of soil both in dry and rainy seasons. In the rainy season, soil with any physical obstructions gets soaking wet suffocating the roots. While in the dry season, the volume of available water is reduced from the roots of a plant; as a results root system of a plant is deepening into the soil for searching available water.

Interrelationship Atmospheric Demand: Soil Water Supply

The soil is a dynamic reservoir for available water for growth and development of plants. It is controlling the rate of water use by the plants and continuously in coincidence with the atmospheric demand. Since the atmospheric demand for water is directing by solar radiation, air humidity, and wind speed. Under situation C, in which ECA (by the evaporation from a Class A pan (ECA)) > 7.5 mm day−1 (high demand); the plants do not manage to extract water at a rate compatible to its needs even under the available soil water, as a result, to avoid drying of the leaves plants temporary close the stomata. Such a condition usually takes place at the hottest hours of the day.

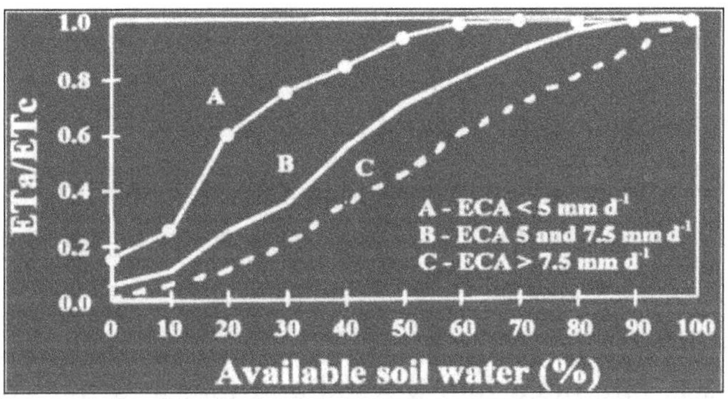

Relationship between corn relative evapotranspiration (ETa/ETc) and soil available water and atmospheric demand expressed by ECA.

Surface Runoff

Runoff is the quantity of water that is discharged ("runs off") from a drainage basin during a given time period. Runoff data may be presented as volumes in acre-feet, as mean discharges per unit of drainage area in cubic feet per second per square mile, or as depths of water on the drainage basin in inches. It is measured by establishing stream gauges at selected places of the river courses. The term runoff refers to the overland flow of water, after every rainfall or snowmelt. The overland flow starts when the rate of rainfall is greater than the rate of infiltration of the soil and increase in the amount of slope. Initially, Runoff starts as small streams and the water gets added from many such streams. Finally, all of these reach and confluence with a lake or stream or directly with seas. The volume of water leaving through a river is known as river discharge. It is considered as precipitation returning to the sea. Stream flow is the discharge that occurs in a natural channel. Although the term "discharge" can be applied to the flow of a canal, the word "stream flow" uniquely describes the discharge in a surface stream course. The term "stream flow" is more general than "runoff" as stream flow may be applied to discharge whether or not it is affected by diversion or regulation.

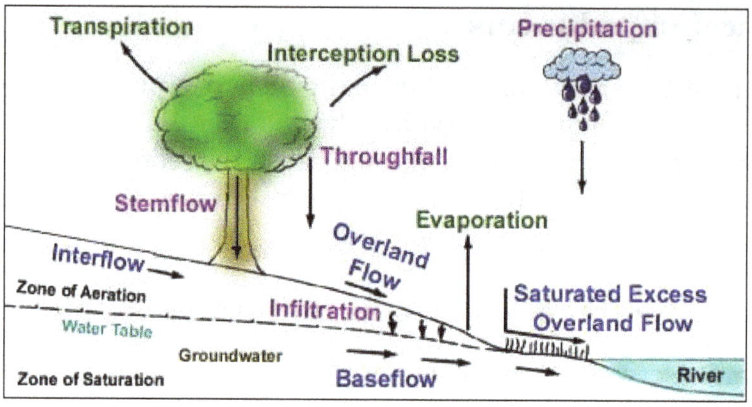

Surface Detention/Detention Storage

The amount of water on the land surface in transit towards stream channels is called detention storage/surface detention.

Types of Runoff

- Surface Runoff: That portion of rainfall which enters the stream immediatcly after the rainfall. It occurs when all loses is satisfied and rainfall is still continued and rate of rainfall [intensity] in greater than infiltration rate.
- Sub-Surface Runoff: That part of rainfall which first leaches into the soil and moves laterally without joining the water table, to the stream, rivers or ocean is known as sub-surface runoff. It is usually referred is inter-flow.
- Base Flow: It is delayed flow defined as that part of rainfall, which after falling on the ground the surface, infiltrated into the soil and meets to the water table and flow the streams, ocean etc. The movement of water in this is very slow. Therefore it is also referred a delayed runoff.

  Total runoff = Surface runoff + GW Base flow.

Urban Runoff

Urban runoff is surface runoff of rainwater created by urbanization. This runoff is a major source of flooding and water pollution in urban communities worldwide. Impervious surfaces (roads, parking lots and sidewalks) are constructed during land development. During rain storms and other precipitation events, these surfaces (built from materials such as asphalt and concrete), along with rooftops, carry polluted storm water to storm drains, instead of allowing the water to percolate through soil.

The Surface Runoff Process

As the rain continues, water reaching the ground surface infiltrates into the soil until it reaches a stage where the rate of rainfall (intensity) exceeds the infiltration capacity of the soil. Thereafter, surface puddles, ditches, and other depressions are filled with water (depression storage), and after that overland flow as runoff is generated. The process of runoff generation continues as long as the rainfall intensity exceeds the actual infiltration capacity of the soil but it stops as soon as the rate of rainfall drops below the actual rate of infiltration.

Factors Controlling Runoff

The flow of any stream is determined by two major groups of factors. The first set belongs to the geomorphological factors of the drainage basin. The second set of factors depends on the climatological variables.

Meteorological factors affecting runoff:

- Type of precipitation (rain, snow, sleet, etc.),
- Rainfall intensity,
- Rainfall amount,
- Rainfall duration,
- Distribution of rainfall over the watersheds,
- Direction of storm movement,
- Antecedent precipitation and resulting soil moisture,
- Other meteorological and climatic conditions that affect evapotranspiration, such as temperature, wind, relative humidity, and season.

Physical characteristics affecting runoff:

- Land use,
- Vegetation,
- Soil type,
- Drainage area,
- Basin shape,
- Elevation,
- Slope,
- Topography,
- Direction of orientation,
- Drainage network patterns,
- A pond, lakes, reservoirs, sinks, etc. in the basin, which prevent or alter runoff from continuing downstream.

The climatological factors are:

- Rainfall – Intensity and Type,
- Duration of Rainfall,
- Distribution of Rainfall,
- Direction of Storm Movement,
- Soil Moisture Conditions.

The geomorphological factors include land use land cover, type of soil, area, shape, elevation, slope, network of drainages and indirect influences for runoff.

Climate Factors

Rainfall Characteristics

- Types of Precipitation: It has great effect on the runoff. E.g. A precipitation which occurs in the form of rainfall starts immediately as surface runoff depending upon rainfall intensity while precipitation in the form of snow does not result in surface runoff.
- Rainfall Intensity: If the rainfall intensity is greater than infiltration rate of soil then runoff starts immediately after rainfall. While in case of low rainfall intensity runoff starts later. Thus high intensities of rainfall yield higher runoff.
- Duration of Rainfall: It is directly related to the volume of runoff because infiltration rate of soil decreases with duration of rainfall. Therefore medium intensity rainfall even results in considerable amount of runoff if duration is longer.
- Rainfall Distribution: Runoff from a watershed depends very much on the distribution of rainfall. It is also expressed as "distribution coefficient". Near the outlet of watershed, runoff will be more.
- Direction of Prevailing Wind: If the direction of prevailing wind is same as drainage system, it results in peak low. A storm moving in the direction of stream slope produce a higher peak in shorter period of time than a storm moving in opposite direction.
- Other Climate Factors: Other factors such as temperature wind velocity, relative humidity, annual rainfall etc. affect the water losses from watershed area.

Physiographic Factors

- Size of Watershed: A large watershed takes longer time for draining the runoff to outlet than smaller watershed and vice versa.
- Shape of Watershed: Runoff is greatly affected by shape of watershed. Shape of watershed is generally expressed by the term "form factor" and "compactness coefficient".
    - Form Factor = Ratio of average width to axial length of watershed.

   - Compactness Coefficient: Ratio off perimeter of watershed to circumference of circle whose area is equal to area of watershed.
   - Two Types of Shape: Fan shape [tends to produce higher runoff very early]. Fern shape [tend to produced less runoff].
- Slope of Watershed: It has complex effect. It controls the time of overland flow and time of concentration of rainfall. E.g. sloppy watershed results in greater runoff due to greater runoff velocity and vice-versa.
- Orientation of Watershed: This affects the evaporation and transpiration losses from the area. The north or south orientation, affects the time of melting of collected snow.
- Land Use: Land use and land management practices have great effect on the runoff yield. E.g. an area with forest cover or thick layer of mulch of leaves and grasses contribute less runoff because water is absorbed more into soil.
- Soil Moisture: Magnitude of runoff yield depends upon the initial moisture present in soil at the time of rainfall. If the rain occurs after a long dry spell then infiltration rate is more, hence it contributes less runoff.
- Soil Type: In filtration rate vary with type of soil. So runoff is great affected by soil type.
- Topographic Characteristics: It includes those topographic features which affects the run-off. Undulating land has greater runoff than flat land.
- Drainage Density: It is defined as the ratio of the total channel length [L] in the watershed to total watershed area [A]. Greater drainage density gives more runoff. Drainage density = L/A.
- Storage Characteristics:
   - Depressions,
   - Ponds, lakes and pools,
   - Stream,
   - Channels,
   - Check dams in gullies,
   - Upstream reservoirs or tanks,
   - Ground water storage in deposits/aquifers.

Condensation

The sun's radiant energy evaporates water from surface water resources, including seas and oceans. This water rises upwards as water vapour and reaches the atmosphere. The process of evaporation continues until the air becomes fully saturated with maximum amount of moisture. It is called as Saturation Humidity. It is directly proportional to the temperature of the air. It is expressed in grams/ cubic metre of air. At 0 degree C, humidity of air is 4.874 gm/ cu.m and at

30 degree C it is 30.38 gm/cu.m. Relative humidity is a term used to denote the ratio of the water vapor content of the air to the water vapor capacity of the air. The term absolute humidity is used to denote the number of grams of water per cu.m of air. If the air mass reaches the dew point, then the condensation process begins. Humidity is measured using hygrometers. Through the process of condensation clouds are formed and moved towards the land frontiers.

Condensation in the atmosphere takes place around small particles of substances present in the air, which have the affinity for water. These form the hygroscopic nuclei surrounding the condensation nuclei. These condensed water vapour masses get precipitated the form of water molecules. Precipitation is the process by which this transformation happens in the atmosphere, under certain specific conditions.

Condensation depends on several factors. The notable areas are:

- Adiabatic cooling,
- Mixture of two air masses of different temperatures,
- Contact cooling,
- Radiational cooling.

Among these, the most important condensation process is the adiabatic cooling which leads to all kinds of precipitation. Condensation leads to cloud formation. Condensed clouds of smaller sizes may quickly disappear on a hot day.

Cloud condensation nuclei, which vary in size from 10^{-5} to 10^{-1} mm, are required to condense water vapour at dew point. Typical cloud condensation nuclei are meteoric dust, windblown clay and silt, volcanic material, sea salt and combustion products. Natural concentrations of cloud condensation nuclei vary from 100 to 300 cm^{-3}, but may locally become 10 to 100 times larger due to human activities.

When a vapour condenses, heat is released. Condensation also leads to the formation of fog. Before precipitation to happen, the cloud droplets sized from 0.001 to 0.2 mm have to grow to 0.4 to 4 mm in diameter to reach a fall velocity that exceeds the rate of uplift. Cloud droplets may grow at temperatures above 0 degree Centigrade. The droplet collision occurs due to differences in fall and lift velocities caused by the differences in droplet size. The world's water vapour is equal to about 2.5 cm thickness, on an average, if it is distributed all over the surface of the globe.

Infiltration

Infiltration is the downward percolation of rainwater (or) snow melt water into the soil horizons. The downward movement of water happens in the top soil layer, especially through the smaller pore spaces present in the soils. Infiltration is governed by two forces as gravity and capillary action. The smaller pores offer greater resistance to gravity; very small pores pull water through capillary action in addition to and even against the force of gravity. Tensiometer is used for the measurements of soil water properties.

After every rain, the zone of soil moisture obtains the first part of the rainfall. The rate of percolation is known as infiltration rate. It is measured using infiltrometers or lysimeters. Infiltration rates are measured by conducting infiltration tests. The rate of infiltration varies from soil to soil. It depends on the hydrologic properties of soils, like porosity and permeability. The process is also known as percolation. The percolating rainwater ultimately reaches the groundwater zone. The process of infiltration happens only when there is space available for addition of water within the soil surface. This depends on the porosity of the soil and the rate at which previously infiltrated water can move away from the surface through the soil.

The maximum rate that water can enter into a soil layer in a given subsurface condition is known as the infiltration capacity. Infiltration rate in soil science is a measure of the rate at which a particular soil is able to absorb rainfall or irrigation. It is measured in inches per hour or millimeters per hour. The rate decreases as the soil becomes saturated. If the precipitation rate exceeds the infiltration rate, runoff will usually occur unless there is some physical barrier.

Factors Affecting Infiltration

- Precipitation: The greatest factor controlling infiltration is the amount and characteristics (intensity, duration, etc.) of precipitation that falls as rain or snow. Precipitation that infiltrates into the ground often seeps into streambeds over an extended period of time, thus a stream will often continue to flow when it hasn't rained for a long time and where there is no direct runoff from recent precipitation.
- Base flow: To varying degrees, the water in streams has a sustained flow, even during periods of lack of rain. Much of this "base flow" in streams comes from groundwater seeping into the bed and banks of the stream.
- Soil characteristics: Some soils, such as clays, absorb less water at a slower rate than sandy soils. Soils absorbing less water result in more runoff overland into streams.
- Soil saturation: Like a wet sponge, soil already saturated from previous rainfall can't absorb much more, thus more rainfall will become surface runoff.
- Land cover: Some land covers have a great impact on infiltration and rainfall runoff. Vegetation can slow the movement of runoff, allowing more time for it to seep into the ground. Impervious surfaces, such as parking lots, roads, and developments, act as a "fast lane" for rainfall - right into storm drains that drain directly into streams. Agriculture and the tillage of land also change the infiltration patterns of a landscape. Water that, in natural conditions, infiltrated directly into soil now runs off into streams.
- Slope of the land: Water falling on steeply-sloped land runs off more quickly and infiltrates less than water falling on flat land.
- Evapotranspiration: Some infiltration stays near the land surface, which is where plants put down their roots. Plants need this shallow groundwater to grow, and, by the process of evapotranspiration, water is moved back into the atmosphere.

Sublimation

Sublimation is the process in which (referring to the water cycle) ice or snow goes from a solid to a gas without becoming a liquid. Deposition is the opposite of sublimation, where the water vapor becomes ice or snow without becoming water.

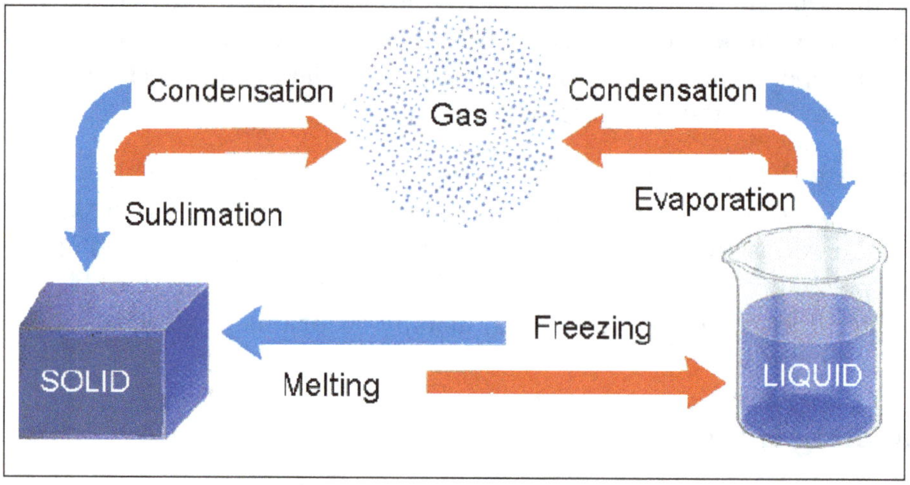

It is difficult to experience sublimation because the process in which gas rises into the air off of something isn't normally clear unless you get close. Sublimation occurs more often with certain weather conditions like low humidity and dry winds. It is more likely for sublimation to occur in higher altitudes.

References

- Evapotranspiration: energyeducation.ca, Retrieved 27, April 2020
- Surface-water-runoff-320331329: researchgate.net, Retrieved 07, June 2020
- Sublimation: thehydrologiccycle.weebly.com, Retrieved 14, May 2020
- Precipitation-is-any-liquid-of-the-global-water-cycle: nationalgeographic.org, Retrieved 07, Jan 2020
- Concept-and-consequence-of-evapotranspiration-for-sustainable-crop-production-in-the-era-of-climate, advanced-evapotranspiration-methods-and-applications: intechopen.com, Retrieved 22, August 2020

3

Aquifer: Types and Properties

An aquifer is the saturated zone beneath the water table. Aquifers are permeable and porous so that the water can travel through. They naturally filter the groundwater by making it pass through small pores. This chapter covers the types and properties of aquifers. The topics elaborated in this chapter will help in gaining a better perspective about the aquifers.

An aquifer refers to a geological formation, which can store and transmit groundwater in sufficient quantity so that the water can be economically utilized from the aquifer. For this to happen the aquifer material should have reasonably good amount of interconnected pores. None indurated sand and gravels are example of some excellent aquifers. Similarly, aquiclude refers to geological formations like clay, which have good volume of pores, but they are not interconnected. Such formations can only store water but they cannot transmit them for economic yield of water. We also have plutonic rocks like granite, diorite etc. having interlocking granular texture. These rocks do not have much voids/pores, thus negligible interconnection of pores. Such formations are referred to as aquifuge, and they neither store nor transmit water. Besides, all these we have aquitards like clayey silt, which may have sufficient volume of pores, but limited interconnectedness of pores. Thus, they may transmit water only in certain directions.

Now we know that the water bearing and transmitting capabilities of different geological formations are not same. However, it is also a fact that in a region mostly we have a group of different rocks and formations occurring together. Hence, it is rare that for some 300 meters thickness you find only aquifer. In general, when we study formations for considerable thicknesses, we find an aquifer could be both overlain and underlain by aquiclude, aquifuge or aquitard. Such different natural geological setting of an aquifer creates different hydrological regimes.

Types of Aquifer

Confined and Unconfined Aquifer

Confined Aquifers are those bodies of water found accumulating in a permeable rock and are been enclosed by two impermeable rock layers or rock bodies. Confined Aquifers are aquifers that are found to be overlain by a confining rock layer or rock bodies, often made up of clay which might offer some form of protection from surface contamination. The geological barriers which are non-permeable and found exist between the aquifer causes the water within it to be under pressure which is comparatively more than the atmospheric pressure. The presence of fractures, or cracks in bedrocks is also capable of bearing water in large openings within bedrocks dissolving some of the rock and accounts for high yields of well in karst terrain counties like Augusta, Bath within

Virginia. Groundwater flow through aquifers is either vertically or horizontally at rates often influenced by gravity and geological formations in these areas.

Confined aquifers could also be referred to as "Artesian aquifers" which could be found most above the base of confined rock layers. Punctured wells deriving their sources from artesian aquifers have fluctuation in their water levels due more to pressure change than quantity of stored water. The punctured well serve more as conduits for water transmission from replenishing areas to natural or artificial final points. In terms of storativity, confined aquifers have very low storativity values of 0.01 to 0.0001.

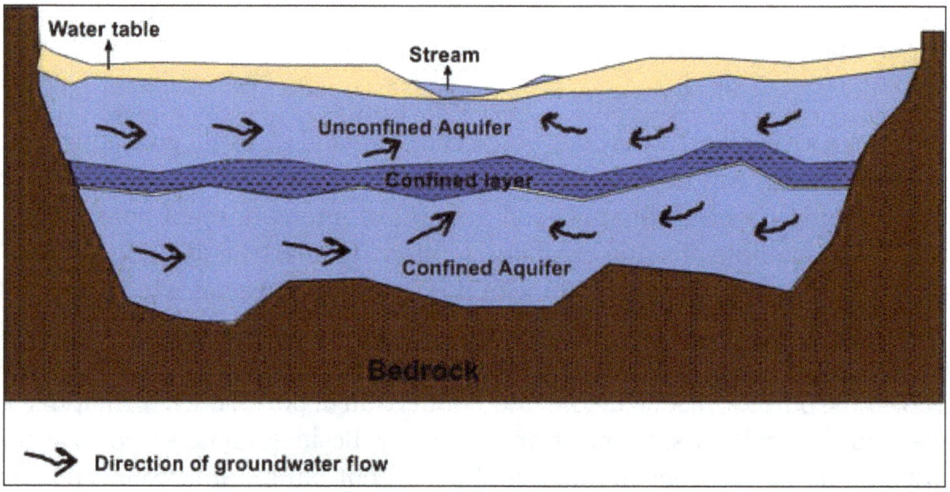

Schematic cross-section of aquifer types.

The hydrogeological basin or the groundwater basin is delineated by its physical limits, which in many cases correspond to the extension of the natural recharge area. The natural recharge occurs mainly by downward percolation through the unsaturated layers under the ground surface, or/and by vertical fluxes (leakage) through the saturated semi-pervious layers called aquitards, from underlying or overlying aquifers with a higher hydraulic head. In order to evaluate the groundwater resources the best way is to consider the whole aquifer. The model extension corresponds to the physical limits of the aquifer, representing the extension in plane of the permeable, hydraulically connected, formations. In some cases, like when determining the effect of hydrotechnical works (dams and reservoirs, irrigation systems, navigable canals or water diversions, huge excavations etc.) a portion of the aquifer can be modelled; this approach is correct only when the internal stresses do not transmit effects interacting with the borders. Pollutants transport is usually modelled for limited areas, representing sub-models of the regional model characterizing the whole aquifer.

Natural Recharge of the Aquifers

Unconfined Aquifers

Natural recharge of the unconfined aquifers is mainly due to the downward seepage (or percolation) through the unsaturated zone of the excess water over passing the field capacity of the soil. Recharge can also occur through upward seepage (leakage) from underlying aquifers.

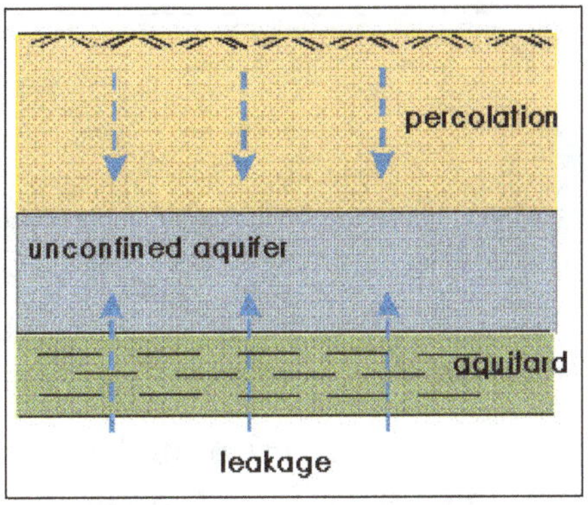

Recharge of the unconfined aquifers.

In the irrigated areas, important quantities of water currently exceeding the field capacity percolate. The losses from the water supply network or the sewage system in urban areas are another source of aquifers recharge.

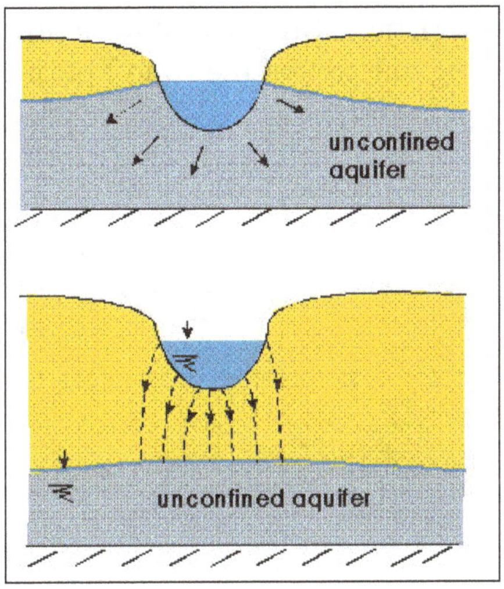

Unconfined aquifers gaining water from the river.

The rivers generally drain the aquifers, but in some cases they can lose water feeding the aquifers; this occurs when the groundwater level is at a lower value than the river water level or during floods, when the water flows in the floodplain. In the arid countries, some rivers infiltrate totally in the bed river if formed by sand or gravel.

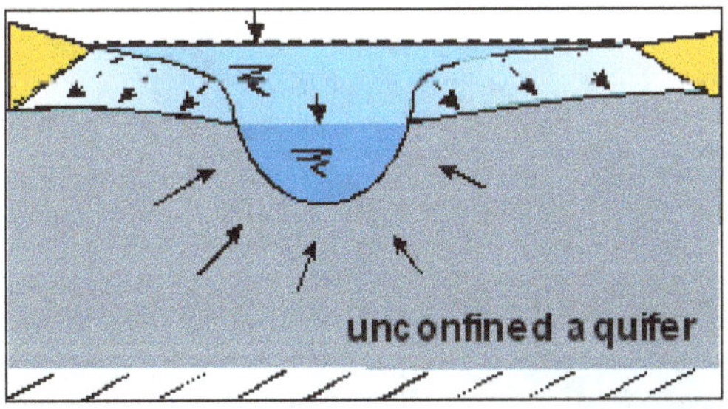

Unconfined aquifers gaining water during floods.

When two rivers drain an unconfined aquifer, an underground water divide corresponding to the crest of the water table separates the drainage areas of each river.

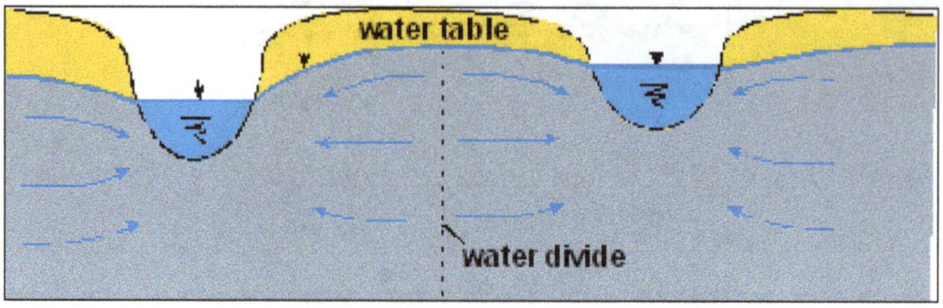

Water divides of an unconfined aquifer.

Many unconfined aquifers used for water supply are located in the alluvial fans, which have a great thickness and contain important volumes of water; in the same time, due to the high values of the hydraulic conductivity, they can transmit the water fluxes quite rapidly for a porous medium.

Confined Aquifers

A regional confined aquifer is directly recharged by precipitation in the area where the aquifer crops out, having the same characteristics as an unconfined aquifer. Another source of recharge is the infiltration in the same cropping out area of the runoff born on the hillslopes during rains or snowmelt.

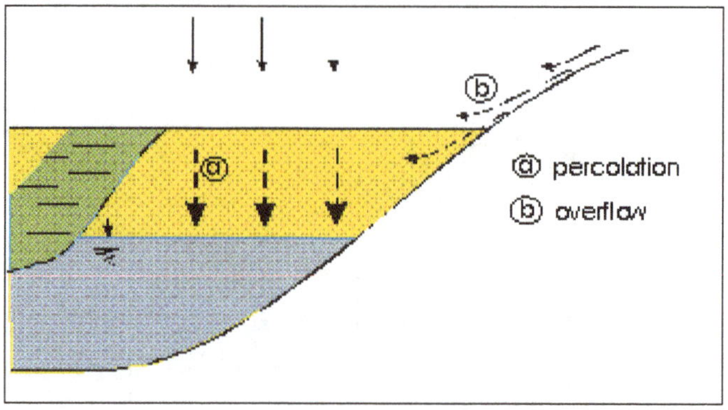

Recharge of a confined aquifer in the cropping out area.

These unconfined aquifers in the direct recharge area follow the bedrock deepening, being covered by impervious or semi-pervious layers and becoming confined aquifers.

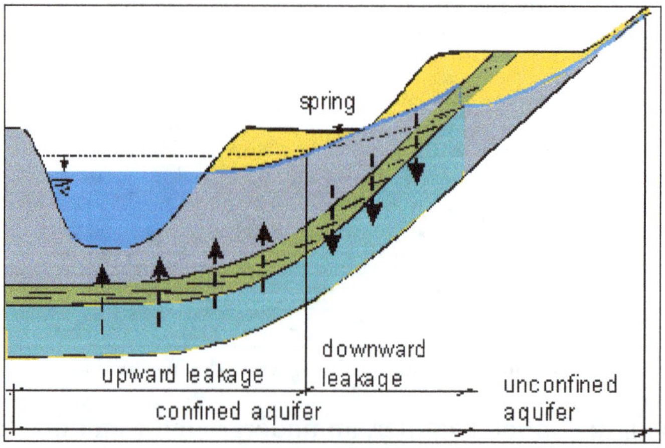

Fluxes in a confined aquifer.

Interesting relations are established in this latter case between the confined aquifer and the overlying unconfined aquifer. According to the position of the water head, the vertical fluxes between the two layers are directed downward or upward.

Thus the confined aquifers are recharged by direct infiltration and percolation or by leakage. In the case of a multi-strata aquifer, depending on the value of the water abstraction, a confined layer can be recharged by vertical fluxes coming from the overlying and underlying aquifers.

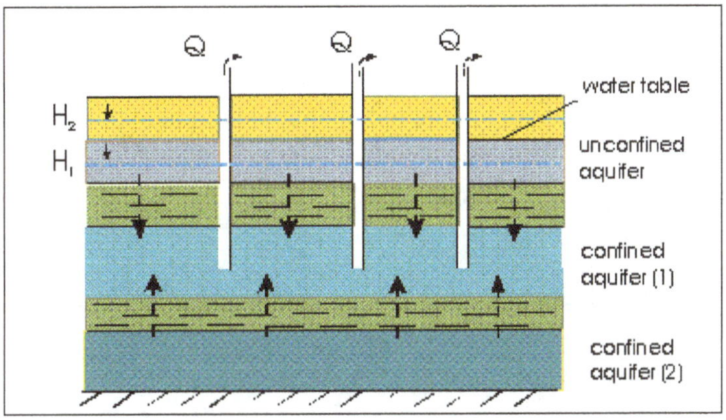

Leakage recharge of a confined aquifer.

The confined aquifer denoted by 1 is characterized by the hydraulic head H_1, where H_2 is the hydraulic head of the second confined aquifer; thus there will be a flux directed from the confined aquifer no. 2 toward the aquifer no. 1. At the same time, the hydraulic head H_1 is under the water table of the unconfined aquifer, which explains the downward flux from the unconfined aquifer to the first confined aquifer.

Discharge of the Aquifers

Unconfined Aquifers

The unconfined aquifers supply with water the permanent river network through the basic flow; thus, the rivers represent the main discharge zones of the unconfined aquifers.

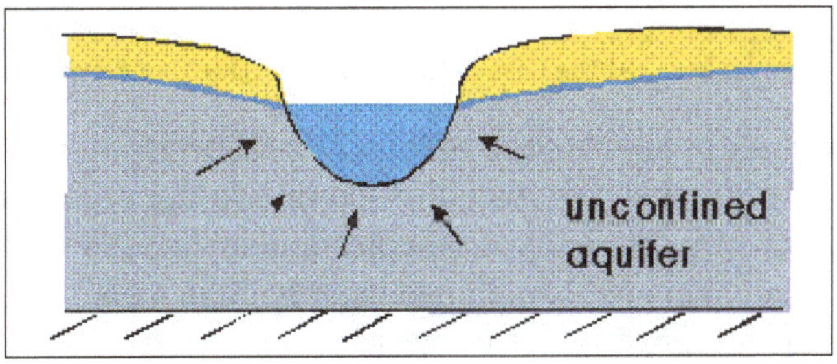

Discharge of an unconfined aquifer in the river.

There are still some cases when the rivers lose water in the aquifers. In the temperate area, this kind of connection occurs only on some sectors of the rivers, the same river losing or gaining water along its course. For instance, a mountainous river loses water in the aquifer when it enters the piedmont plain, but in the downstream sector because the riverbed is deepened, the same river drains the aquifer.

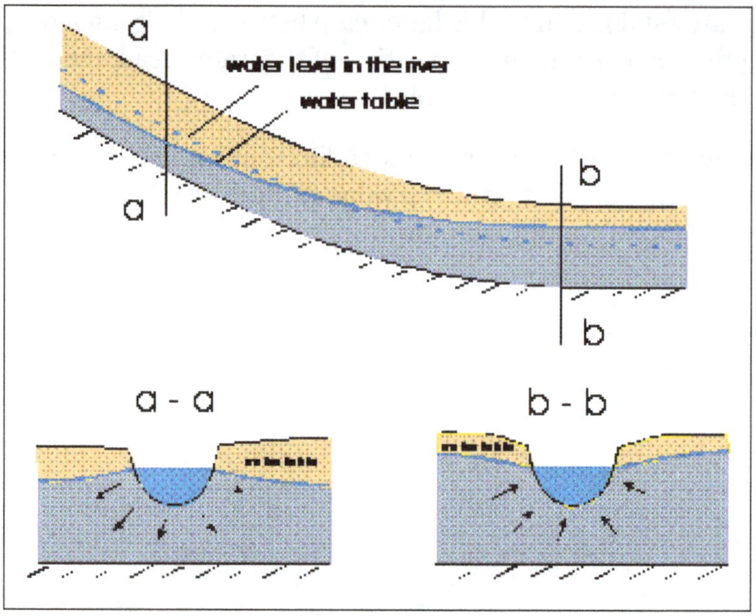

Relationship surface water-groundwater along the river.

If the water table intersects the ground surface a spring can occur. These springs are generally present at the terraces base; if an impervious layer does not allow the normal percolation and a perched aquifer is formed, a spring can flow at a higher elevation.

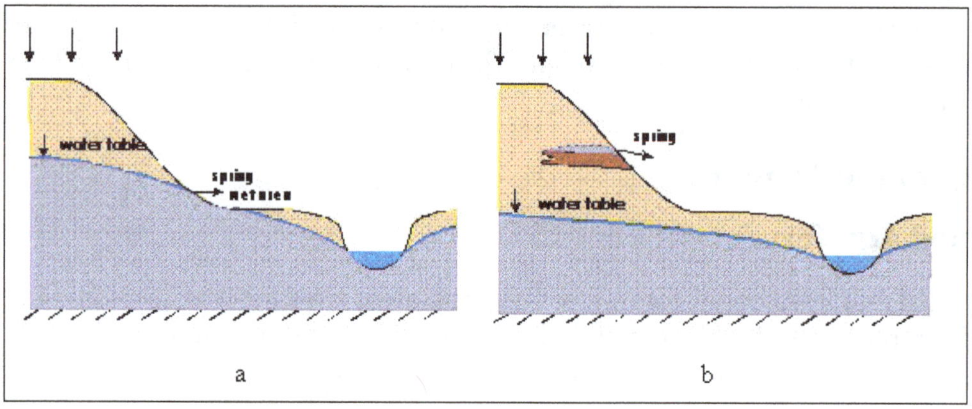

Spring occurrence: a) At the base of the terrace; b) Due to a perched aquifer.

Generally, the discharge areas are quite close to the ground surface. Water can be lost not only by concentrated discharge like streams, but also diffused due to the vegetation, through the evapotranspiration, or directly by evaporation from the ground surface, the superficial soil being supplied with water by capillarity. Very well developed vegetation in a plain area or the presence of the salt at the ground surface can indicate a discharge area.

The natural depressions drain also the groundwater; if the discharge is small compared with the evapotranspiration of the affected area, only a wet zone is present, and sometimes with a temporary character. If the discharge and the natural precipitations are more important than the evapotranspiration, a lake can be formed.

The sea, being at a lower elevation, represents the discharge area for the neighbouring unconfined aquifers. The discharge is partially above the sea level by evaporation or diffused springs and partially under the water level.

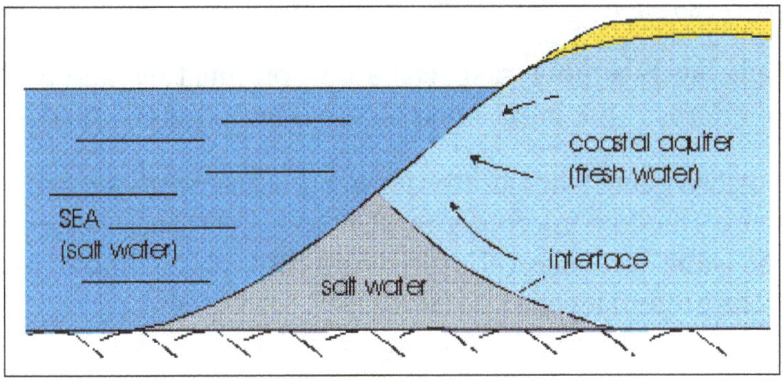

Coastal aquifer.

A delicate equilibrium in the coastal areas is established between the salt water and the fresh water. If the abstracted fresh water overpasses the natural discharge into the sea, the salt-water intrusion takes place, and the quality of the water does not correspond anymore.

Confined Aquifers

The confined aquifers are less influenced by the topography than the unconfined aquifers. The discharge of the confined aquifers occurs usually by leakage, either in an unconfined aquifer, or in a confined aquifer, having a smaller hydraulic head. In some cases, the discharge of a confined aquifer is made through an outcrop area, situated in a plain region at the opposite side to the recharge zone.

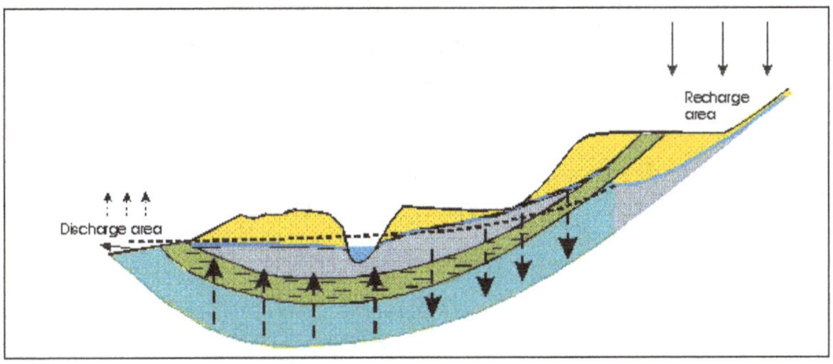

The discharge of a confined aquifer.

Coastal confined aquifers have as discharge area the sea, either as diffused discharge or as springs under the sea water level.

Water Table and Piezometric (Potentiometric) Surface

Both water table for unconfined aquifers and piezometric or potentiometric surface for the confined aquifers can be represented in the same way as the surface topography, using lines of equal elevation. The datum is usually the sea level and the contour lines represent the intersection of the water surface with horizontal planes, equally distanced on the vertical. The data used for constructing the contour lines are the water levels elevations measured in wells in a regime non-influenced by the water abstraction. That means that during the field campaign or when making the regular level measurement the wells used for water supply must be shut off, the observer having to wait until the water level is stabilised. This level is termed as static or non-pumping level, to distinguish it from the dynamic level occurring during the water abstraction.

In the case of a multiple layers aquifer, a special attention must be paid to the water levels measured in the aquifer of interest; that means that all the other aquifers must be cased off.

The contour lines are drawn using a linear interpolation. This technique can create problems if the ground surface is convex. Due to a micro-depression area a wet zone is formed, while using a linear interpolation of the water level elevations between the two neighbouring wells the water level would be higher than the ground level.

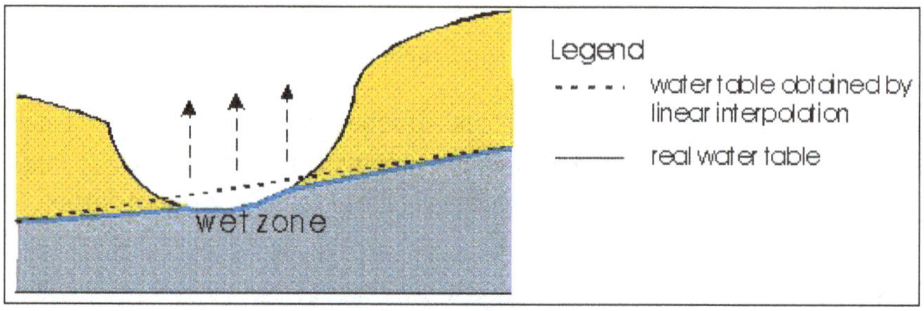

Limits of the linear interpolation.

Contours of equal groundwater elevations must be drawn very carefully and with a perfect knowledge of the field situation; they can provide a future user with right or wrong information that will be used for mathematical modelling purposes. For instance, the contour lines will not be the same if a river is hydraulically connected with the aquifer or is separated by impervious layers. In the same time, according to the orientation of the aquifer-river fluxes, the shape of the contour lines will be different. The flowlines, which are perpendicular on the contour lines, indicate the sense of the water fluxes. To draw the contour lines in this case, water levels measured in a battery of wells perpendicular to the river are necessary. In the absence of such measurements, the way the contour lines are drawn is a problem of interpretation.

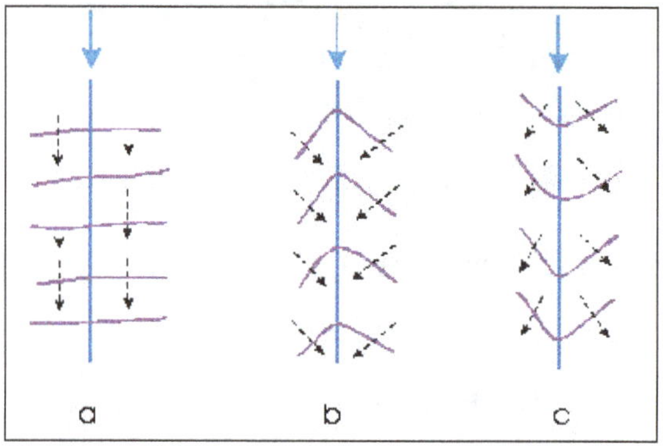

Using contour lines to interpret the hydraulic connection between aquifer and river: (a) No hydraulic connection between river and aquifer. (b) The river draine the aquifer. (c) The aquifer gains water from the river.

The contour lines provide also qualitative information concerning the hydraulic conductivities contrasts. If in the flow sense, the contour lines are distanced, being followed by close contour lines, in the first zone the hydraulic conductivities are greater than in the second zone. The hydraulic gradients being greater in the second region than in the first area, the energy losses are greater, meaning a greater hydraulic resistance or, what is the same thing, smaller hydraulic conductivities than in the first area.

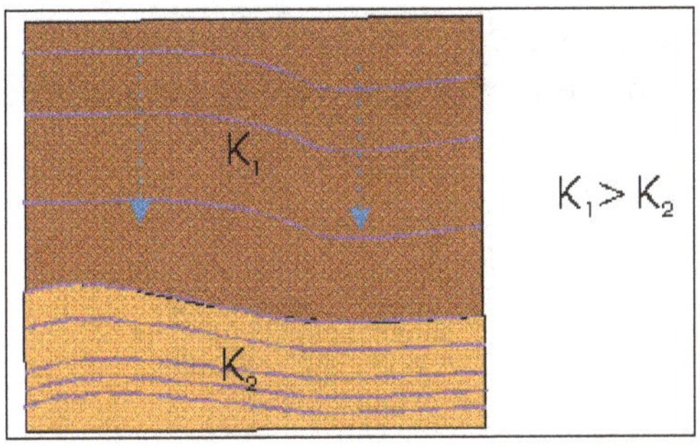

Hydraulic gradients and hydraulic conductivities.

In some cases, the recharge or the discharge area are characterized by specific contour liners. The recharge area is characterized by higher elevations than the rest of the aquifer, and the flow lines diverge radially; an inverse situation occurs in the discharge area, where the flow lines converge radially.

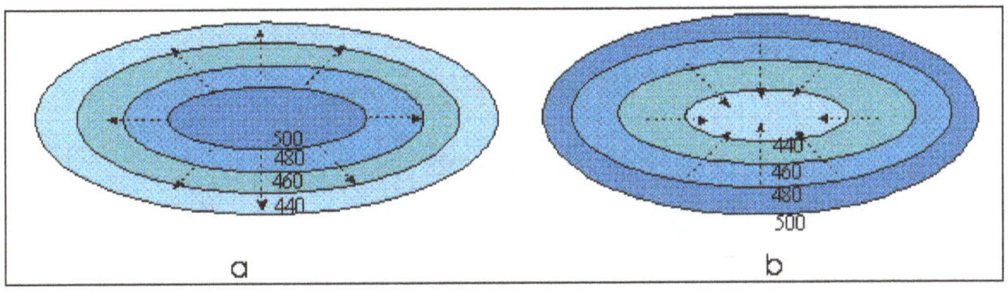

Special shapes of contour lines in recharge and discharge area.

Still, in most of the cases, the recharge and the discharge area are defined only by higher, respectively lower elevations values than in the rest of the aquifer. The contour lines are perpendicular on the impervious boundaries.

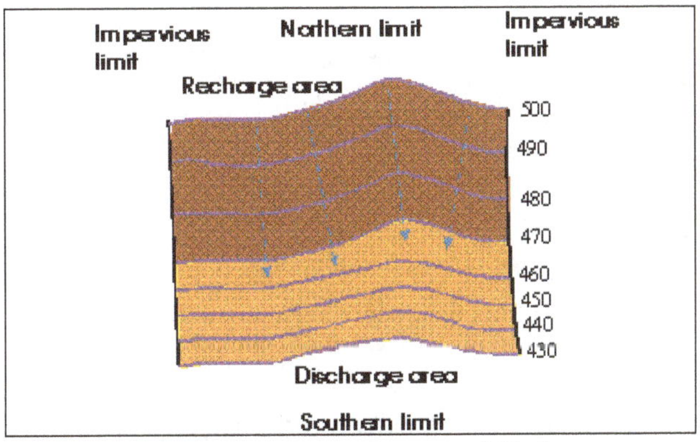

Common contour lines in recharge and discharge area.

Interesting shapes of the contour lines are obtained in the case of the water works; in the vicinity of each well a depression cone is created. According to the distance between wells, there are interferences in drawdown (cross-section A-A). For a homogeneous medium, at the same discharge the drawdown is symmetrical due to the symmetry of the problem; the central well has a greater value due to the influence of all the other wells. At the same time, the upstream influence zone is greater than the downstream zone (cross-section B-B).

Water Reserve and Water Resource

The natural reserves, also called potential reserves or total reserves represent the volume of water that can be mobilised at a certain moment. Usually, the mean multi-annual value of the natural reserve is considered, having a different expression for the unconfined, respectively confined aquifers.

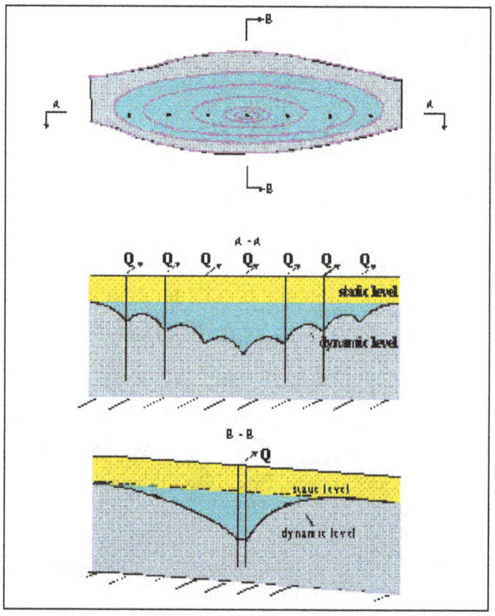

Contour lines and vertical cross-sections through a battery of wells.

Unconfined Aquifers

Considering the effective porosity ne of an elementary volume $d\overline{V}$, the mean natural reserve $\overline{W}$ can be obtained by integration:

$$\overline{W} = \int_A n_e \cdot \overline{H} \cdot dA = \int_A n_e \cdot H_{min} \cdot dA + \int_A n_e \cdot \Delta\overline{H} \cdot dA = W_{min} + \Delta\overline{W}$$

Where,

- A = Plane area of the aquifer.
- $\overline{H}$ = Average value of the saturated depth of the aquifer, computed for a duration T and for the whole area A.
- H_{min} = Minimum or the permanent saturated depth, over the time.
- W_{min} = Permanent or the geological reserves, representing that part of the potential reserves which is not influenced by the natural variation of recharge and discharge values.
- $\Delta\overline{H}$ = Mean value of the water level fluctuations above the permanent saturated depth H_{min} computed for the duration T and for the whole area A.
- $\Delta\overline{W}$ = Mean temporary reserves.

Taking into account that for 2D aquifers H = H(x, y, t), the average value of the saturated depth is obtained through a double integration, in space and in time:

$$\overline{H} = \frac{1}{T}\int_{0}^{T} \overline{H}_A(t)dt$$

$\overline{H}_A(t)$ is the average saturated level on the whole aquifer at the moment t:

$$\overline{H}_A(t) = \frac{1}{A}\int_{A} H(x,y,t)dA$$

A schematic representation for an unconfined aquifer of the groundwater level variation and of the corresponding permanent and temporary reserves is presented in figure.

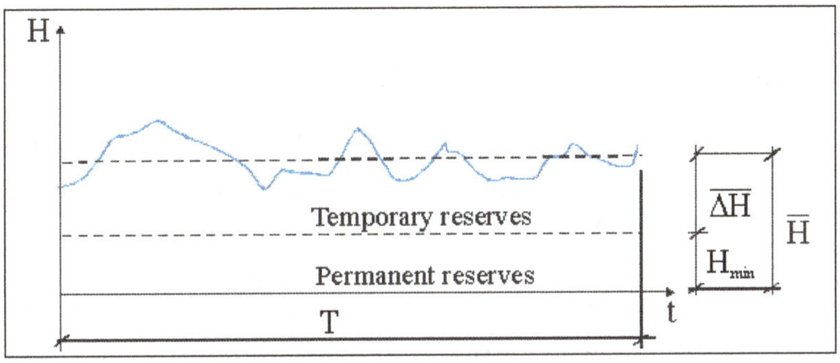

Permanent and temporary reserves.

Confined Aquifers

Considering that the unconfined aquifers always remain under pressure, the mean natural reserve is obtained by a relation similar to $\overline{W} = \int_A n_e \cdot \overline{H} \cdot dA = \int_A n_e \cdot H_{min} \cdot dA + \int_A n_e \cdot \Delta\overline{H} \cdot dA = W_{min} + \Delta\overline{W}$ with the difference that the water is stored in the porous media not only in the voids, but also due to its compressibility:

$$\overline{W} = \int_A (n_e + S_s) \cdot \overline{H} \cdot dA = \int_A n_e \cdot M \cdot dA + \int_A S_S \cdot \Delta\overline{H} \cdot dA = W_o + W_e$$

Where,

- n_e = Effective porosity.
- S_S = Specific storage.
- $\overline{H}$ = Average value of the piezometric level of the aquifer, computed for the duration T and for the whole area A.
- M = Thickness of the confined aquifer.
- $\Delta\overline{H}$ = Mean value of the piezometric level fluctuations above the top of the confined aquifer, computed for the duration T and for the whole area A.

- W_o = Permanent or the geological reserves, corresponding to the effective porosity of the aquifer.
- W_e = Elastic reserves, due to the water and soil compressibility.

For a sustainable management of the aquifers, either unconfined or confined, the permanent or the geological reserve must not be used for water supply purposes. When during exploitation the groundwater levels are situated in a zone corresponding to the geological reserve, the abstraction is greater than the recharge. If this situation continues for a long time, the water levels will decrease continuously with all possible consequences: wells clogging due to increased water access velocity as a result of forcing the wells to abstract the same discharge followed by shortage in water supply, extension of the influence radius and prelevation of water originating from polluted areas, salt water intrusion or even aquifer's depletion. In such cases, urgent measures to rehabilitate the aquifer are necessary; artificial recharge of the aquifer in order to re-establish the initial water levels and/or a drastic limitation of the abstracted water represent solutions to these problems. In order to avoid them, a limited amount of water withdrawal must be allowed.

The American school of hydrogeology defined the safe yield of a groundwater basin as the amount of water that can be withdrawn from it annually without producing an undesired result. By undesired result one understand not only the quantitative aspect (unacceptable drawdown, increased cost of water pumping, wells clogging, depletion of the aquifer), but also qualitative aspects (extension of the polluted areas due to modifying the natural water circulation or increasing natural groundwater gradients towards the water works, salt water intrusion etc.) or legal ones (violation of the existing water rights, especially in the case of trans boundary basins). The yield concept considered in a socio-economic sense within the framework of the optimisation theory led to the optimal yield by selection of the optimal groundwater management scheme from a set of possible alternative schemes.

The French school and some Eastern hydrogeological schools limited the abstracted water to the aquifer resource, representing the quantity or the volume of water that can be withdrawn in a given period of time, taking into account the technical, socio-economic and political constraints.

In natural conditions, the mean multi-annual recharge is equal to the mean multi-annual discharge; if such equilibrium would not exist, the water levels would increase continuously reaching the ground surface or in the other extreme would decrease continuously leading to the aquifer depletion.

According to the groundwater resource philosophy, the water withdrawn from the aquifer will diminish only the natural discharge (base flow of the rivers, lakes or swamps inflow, evapotranspiration), but will not affect in any way the permanent reserve. In fact, the process is more complex, because when withdrawing water above certain values the sense of water fluxes can be changed. For instance, the aquifer instead of losing water in the rivers may be supplied with water from the hydrographic network; in the same time, leakage from underlying or overlying aquifers can be activated.

The following categories of groundwater resources can be distinguished:

1. Natural resource Q_n that represents the discharge of the aquifer in natural conditions.

For an unconfined aquifer, the natural resource is:

$$Q_n = A \cdot q = h \cdot L \cdot K \cdot i = L \cdot T \cdot i$$

Where,

- h = Saturated depth of the aquifer.
- L = Width perpendicular on the flow direction for which the natural discharge is computed.
- T = Aquifer transmissivity.
- i = Hydraulic gradient.

For a confined aquifer, the natural resource is:

$$Q_n = A \cdot q = M \cdot L \cdot K \cdot i$$

Where, M represents the thickness of the aquifer.

2. Supplementary resources Q_S are the discharges that could not be withdrawn before, being the result of the groundwater levels changes induced by water abstraction from the aquifer; new exchanges by leakage or from the river network appear due to the level decrease in the aquifer.

3. Exploitable (or available) resources Q_e represent the discharge that can be withdrawn from the aquifer, maintaining on long term the equilibrium between input and output (including the withdrawn discharge) in a sustainable management, without any danger of aquifer depletion. One can distinguish:

- Budget resources - Exploitable resources whose abstraction is technically possible, economically rational and satisfying the quality criteria for water users.
- Out of budget resources - Exploitable resources whose abstraction is economically irrational in the present due to the reduced amount of water, high cost, bad quality or due to difficult abstraction conditions, but which could be used in the future.

Water Balance

In a very simplified way, the water balance or the hydrologic balance equation for a multi-annual period can be represented like in the figure.

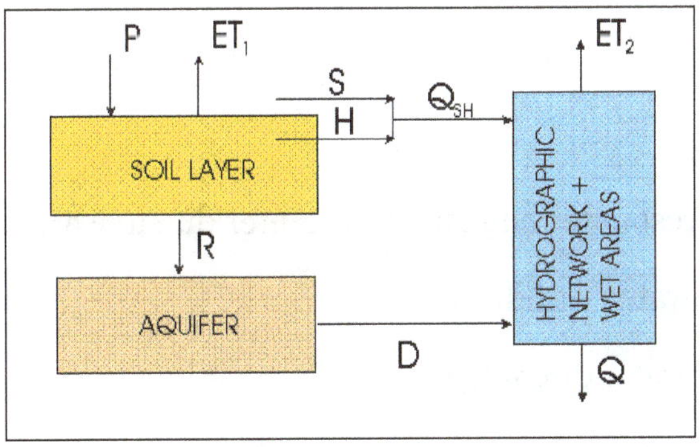

Water balance in an aquifer under natural conditions.

The river network and the wet zones represent the discharge area of the watershed. In natural conditions, the mean multi-annual recharge R of the aquifer is equal to the mean multi-annual groundwater discharge D:

$$R = D$$

If an amount W of water is withdrawn from the aquifer, all the components of the global water balance are more or less modified.

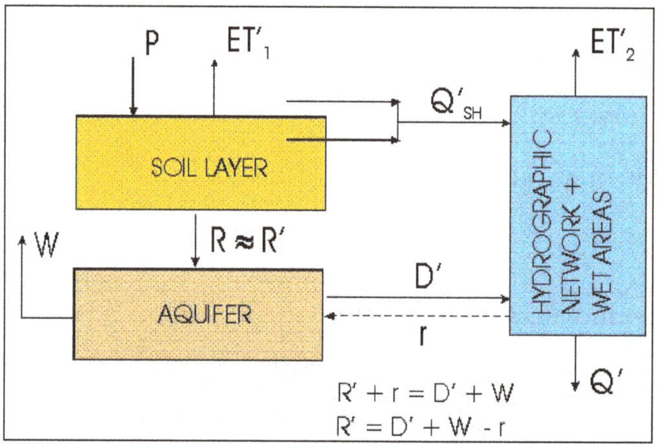

Water balance in an aquifer under exploitation.

The new discharge D' in the rivers and wet zones is smaller than the former value D. The local level in the aquifer in the proximity of water works decreases; a quantity r of water can infiltrate from the river network in the aquifer. At the same time, some modifications in the soil humidity occur, having still a reduced influence on the other components of the hydrologic cycle, like: the surface or the hypodermic runoff, and the evapotranspiration. The multi-annual water balance of the aquifer if water is withdrawn from the aquifer becomes:

$$R' \approx R = D' + W - r$$

This relation is valid only for a multi-annual period, suggesting a steady state regime. For a shorter period (year or month for example) changes of water storage in the aquifer must be introduced. The transient hydrologic budget has the following form:

$$R' = D' + W - r + \frac{dV}{dt}$$

Where,

- R' = The recharge.
- r = Supplementary resources entering the aquifer due to water level drawdown.
- W = Rate of groundwater withdrawal.
- D' = Rate of groundwater discharge.
- dV/dt = Rate of change in storage in the saturated zone.

All the terms in the previous equation, except the withdrawal W, which is relatively constant, are now functions of time. During dry periods the volume of water stocked in the aquifer supplies the withdrawal W. If the dry period is very long, the natural recharge R', the discharge D' and even supplementary resources are null:

$$R' = 0; D' = 0; r = 0$$

Thus, the equation $R' = D' + W - r + \frac{dV}{dt}$ becomes:

$$0 = W + \frac{dV}{dt}$$

Or, after discretization:

$$V_{i+1} = V_i - W \cdot \Delta t$$

The aquifer is acting like a reservoir. Still, the following aspects must be taken into consideration for an unconfined aquifer:

- The aquifer cannot be completely emptied, a minimum prescribed water height being necessary in order to allow the water withdrawal; thus, the groundwater height near the wells must be at least at the upper level of the faints in order to avoid high water access velocities and wells clogging.
- The unconfined aquifers present natural level fluctuations as a result of the precipitations regime; the mean water level must be used when computing the groundwater reserve existing upstream the battery of wells.
- The bottom of the aquifer has a certain slope; due to this fact and to the previous considerations only a part of the volume stocked in the aquifer is effectively available; as a safety measure, one can take into account only the volume of the aquifer situated upstream the wells. Ignoring these facts it is impossible to provide the prescribed groundwater withdrawal with all the consequences for the water users.

In a transient hydrologic budget the evaluation of the time dependent variables like the discharge D’ or the supplementary attracted resources r is difficult and uncertain. Their values depend on the evolution of the water levels in the rivers, which vary in time and space. More than that, due to the differences in soil nature or the land use, the natural recharge varies with the spatial coordinates. Only for a small aquifer one can accept a unique value of the natural recharge; in most of the cases the aquifer has to be divided into several zones characterized by different values of effective infiltration. All these considerations prove that a global model does not represent properly such a complex structure like an aquifer is. The equations $R' \approx R = D' + W - r$ or $R' = D' + W - r + \frac{dV}{dt}$ are only rough approximations of the water budget.

These facts explain the success of the distributed-parameters models based on the areal variation of the hydrogeological parameters and the natural recharge. An aquifer is discretized using a regular or irregular mesh, which defines elements having finite dimensions; the general flow equation is written for each element of discretization resulting thus a budget equation for every element, and implicitly for the whole aquifer. The discharges that intervene in the continuity equation are computed using the Darcy law; recall that the hydraulic gradients involve the computation of the increment of the hydraulic head. As a result, after the discretization of the general flow equation a system of linear equations having as unknowns the hydraulic head values in the center or in the nodes of each discretization element will be obtained. The resulted number of equations is equal to the number of the elements. The general form of this system is:

$$A \cdot H = B$$

Where, A is a matrix depending on the hydraulic conductivities and B is a vector containing the source/sinks terms: natural recharge, withdrawn discharges and the effect of the boundary conditions. The computed groundwater levels H are then used to obtain the water fluxes that input or output the aquifer. A global water balance of the aquifer can be calculated on the basis of these discharges. It is obviously that such a balance is more accurate than the water balance obtained with the global relations. At the same time, the water levels in the vicinity of the water works are obtained with a high degree of accuracy.

Any engineering works that interfere with the aquifer, like dams and reservoirs, navigable canals, tunnels, huge excavations, or any modification in the present recharge-discharge regime having as result the increase of the water percolation (artificial recharge of the aquifer, irrigation systems characterized by important water losses) or the increase of the water withdrawal change the water levels and the water fluxes in the aquifer. The modified water levels can be obtained by solving a system of equation similar to previous equation:

$$A' \cdot H' = B'$$

A new vector of boundary conditions B' that take into consideration the effect of the foreseen interventions replaces the former vector B. In the case of hydraulic structures, some small differences have to be operated in the matrix A, which becomes B'. On the contrary, if the modifications in the aquifer behavior concern only the input or output fluxes without any structural changes, the matrix is invariant. As a result of the modification of the right hand side vector, the solution of the system id modified, becoming H'. If this solution does not locally satisfy some constraints like the minimum acceptable water levels, a new scenario of the aquifer alterations can be imagined. Obviously, solving the system of equations $V_{i+1} = V_i - W \cdot \Delta t$ or $A \cdot H = B$, which involves sometimes thousands of equations and unknowns, is not possible without strong computers. The resulted water levels are in agreement with the equation of continuity; as a result of any simulation, local respectively global water balances are obtained.

Databases and GIS for Handling Information

A large quantity of data is required for regional hydrogeological investigations. Among these data the most important are: the lithology, the wells location, the wells characteristics including the screen position, the history of levels and withdrawals. Other relevant information concerns the topography, soil nature, land use, climatology hydrology and geology of the studied area. All the information is put into a database; Microsoft Access and ORACLE are the most often used environments for creating databases.

The availability of these data determines the understanding of the physical hydrological processes that occur; practically in all the cases the recorded data are not sufficient for the modeling purposes and new data have to be collected through field campaigns. Thus, a special attention must be paid to measurements of new data and achievement of new pumping and tracer tests in order to insure the reliability of the models. The lack of reliable data leads to high uncertainty levels in the model results.

In the last years, due to the necessity of spatially distributed information for the 'process-based' numerical models used in hydrogeology, the Geographic Information Systems (GIS) are largely used. A GIS combines spatial information (geographic coordinates) with attributes for characterizing parameters of any geo-referenced data describing natural environment or man-induced phenomena. GIS can provide a wide choice of functions to access, manage, manipulate and display data.

The practical implementation of all the data in the numerical models can be done using a GIS, by creating direct links with the hydrogeological database. Internal GIS programming languages provides the flexibility needed to link computer codes with GIS applications for such tasks as formatting data files, running numerical model simulations, and post-processing of the results. Representing input and output data through powerful graphical GIS applications linked directly to a predictive numerical model, helps for site characterization, risk assessment and future management strategies.

Purpose of the Groundwater Modeling of Regional Aquifers

The mathematical modelling of regional aquifers has the following purposes:

- To predict the future behaviour of the studied aquifer in response to stress-factors like new pumping, infiltration changes due to global changes, irrigation, and pollutant contaminations.
- To provide information required in order to comply with local regulations.
- To obtain a better understanding of the aquifer system from the geological, hydrogeological and hydrochemical point of view.
- To provide information for the improving of the observation networks and field experiments.

The geologic medium is not an engineering system, where all the characteristics or parameters are deterministically known. 'Physically consistent' models describing the movement of water and contaminants through the geological porous medium are based on the REV (Representative Elementary Volume) approach. The parameters expressing the characteristics of the aquifer (hydraulic conductivity, effective porosity, dispersivities etc.) are 'equivalent' or averaged values on the REV. The hydrogeological parameters are obtained through a calibration process applied for the whole aquifer, based on the best agreement between the measured and the computed levels. Local heterogeneities are taken into account by different values of the chosen parameters. The accuracy of the simulation strongly depends on the scale of the problem.

It has been recently shown that contaminant transport properties of geological media cannot be deduced from laboratory tests. A strong scale effect affects, among others, the dispersivity coefficients. A reliable and quantitative way of determining the aquifer transport properties consists in interpreting two-well injection-pumping tracer tests. A pulse of tracer-labeled water is injected in one well and pumping in a nearby well creates a radially converging flow field. Tracer breakthrough curves are measured in this last well. This information is used to evaluate aquifer transport properties and to calibrate the contaminant transport model. Interpretation by modeling local transport conditions of the aquifer is not an easy task.

For assessing protection zones around main pumping wells, detailed simulations of groundwater transport behavior must be completed involving:

- The local calibration of the model for flow based on measured piezometric levels.
- The local calibration for transport (on the measured breakthrough curves from tracer tests or measured contaminations).
- The simulations and computations of travel times for different injection points.

To resume, the mathematical modeling of a regional aquifer implies the following steps:

- Collecting all existing data.
- Building and development of data-bases.
- New measurements campaigns.
- Regional modeling.
- Groundwater quality studies in local zones.

The mathematical models make possible any simulation to optimize the management and the protection of the groundwater resources. The results of these simulations make possible the sustainable use of groundwater resource, representing a physically and environmentally based decision support system for decision makers.

Semi-confined

A semi-confined aquifer is one that is either overlain or underlain by a semi pervious material, which is mostly an aquitard. These aquifers are also referred to as leaky confined aquifers. A schematic drawing of such aquifer is shown as figure. In an assumed three-layer system, the top layer is semi pervious and the bottom layer is impervious. Such leaky confined aquifers or semi confined aquifers has provision for water movement through the top semi pervious layers.

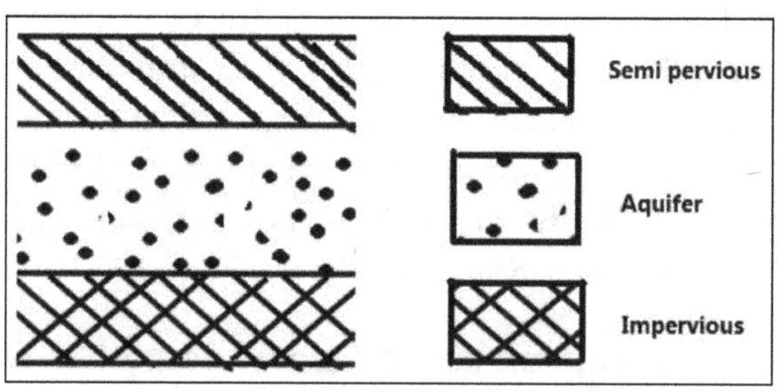

A schematic drawing of a semi confined aquifer.

Perched Aquifer

It often happens that we have a thick unconfined aquifer with some small clay layer at shallow depths. During the process of aquifer saturation through infiltration and percolation by local in-situ recharge, these clay layers create local saturation at shallow depth in the vadose zone much above the water table. Such shallow depth saturation in the vadose zone much above the real saturation zone is often referred to as perched groundwater and the related aquifer is referred as perched aquifer. It is often seen that shallow tube wells tap the perched aquifers. However, in professional practice it is always advisable to have water wells through the perched aquifer tapping the groundwater below the water table zone.

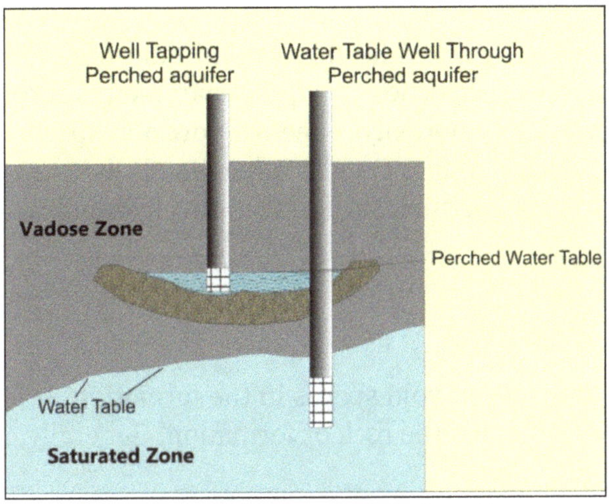

A schematic drawing of perched aquifer.

Leaky Aquifers

If an aquifer (confined aquifer or unconfined aquifer) loses or gains water through adjacent semi-permeable layers, it is called a 'leaky aquifer'. Therefore, the terms 'leaky confined aquifer' and 'leaky unconfined aquifer' are widely used depending on whether the leaky aquifer is confined or unconfined. However, the case of 'leaky confined aquifer' has been mostly dealt with by the groundwater experts. This is why; the term 'semi-confined aquifer' is sometimes used to denote a 'leaky aquifer'. The term 'nonleaky' is also used to describe the status of a confined or unconfined aquifer, such as 'nonleaky unconfined aquifers' and 'nonleaky confined aquifers'. In reality, ideal confined aquifers or ideal unconfined aquifers occur less frequently than do leaky aquifers.

Perched Aquifers

A perched aquifer is a special type of an unconfined aquifer, in which water exists under water-table conditions. Therefore, the upper boundary of this aquifer is called 'perched water table'. Perched aquifer always exists in the vadose zone above an unconfined aquifer or a confined aquifer when a low-permeability layer impedes the downward movement of water above it. Perched aquifers have generally very limited areal extent and they may not have sufficient storage to support significant well production. Therefore, perched aquifers are not the target of a groundwater exploration. However, perched aquifers can support shallow dug wells, thereby can provide water supply to a small community for a limited time period.

It should be noted that hydraulically single aquifers seldom exist in nature. An aquifer is generally part of a system of two or more aquifers, which is more complex. Aquifer thickness, hydraulic head, Darcy velocity, seepage velocity, hydraulic conductivity, transmissivity, intrinsic permeability, storage coefficient (specific storage), specific yield, and specific retention are the important hydraulic and hydrogeologic parameters which are used to characterize an aquifer system.

Aquifer Properties

The hydrogeologic factors which govern the storage and fluid-transmitting characteristics of an aquifer system are called 'aquifer properties' or 'aquifer parameters'. Storage-related aquifer properties (or storage parameters) are: porosity, effective porosity, specific retention, specific yield, storage coefficient, and specific storage. On the other hand, fluid-transmission-related aquifer properties (or yield parameters) are: intrinsic permeability, hydraulic conductivity, and transmissivity.

Porosity

Porosity of a formation is measure of void spaces in the formation. It is expressed as ratio of the volume of voids to the total volume of the rock or formation. Generally, it is expressed as percentage. Thus:

$$n = \frac{V_V}{V} \times 100$$

Where,

- n = Porosity.
- V_v = Volume of voids.
- V = Total volume of rock.

A look at previous equation reveals that porosity is ratio of two volumes and does not have any unit, rather it is dimensionless.

An aquifer can have either primary or secondary porosity. The flow chart in figure clearly shows that the primary porosity is formed during genesis of the rock, while secondary porosity is formed after genesis of the rock. The porosity of loose sand is the best example of primary porosity, while porosity imparted to hard rock because of fracturing is an example of secondary porosity.

Primary porosity is a function of sorting, packing, shape and fabric of the grains, while secondary porosity is a function of intensity of fracturing, degree of solution of the hard rock etc. It must be mentioned explicitly that if all the grains are spherical then in such case porosity is not a function of the diameter of the sphere. Thus if you just magnify the grain size while sorting, packing etc. of the grains remain same, and then the porosity remains same.

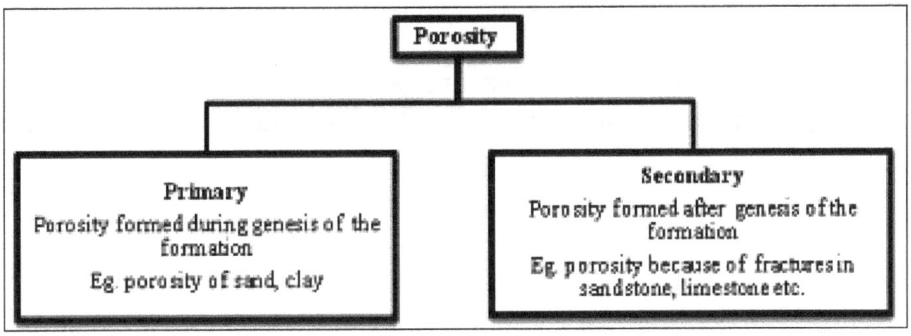

Types of porosity.

Effective Porosity

Effective porosity of a formation refers to the fraction of porosity available for fluid flow. It is measured as a ratio of interconnected pore space/voids available for fluid flow to the total volume of the rock or formation. Generally, it is expressed as percentage. Thus:

$$e = \frac{V_i}{V} \times 100$$

Where,

- e = Effective porosity.
- V_i = Volume of interconnected pore space/voids.
- V = Total volume of rock.

A look at previous equation reveals that effective porosity is ratio of two volumes and does not have any unit, rather it is dimensionless.

Specific Yield

Specific yield of a rock or formation is measured as the ratio of volume of water that after saturation is yielded/drained under influence of gravity to the volume of the rock or formation:

$$S_y = \frac{V_W}{V}$$

Where,

- S_y = Specific yield.
- V_w = Volume of water drain.
- V = Total volume of rock.

Specific yield is ratio of two volumes and does not have any unit, rather it is dimensionless. It is to be noted that specific yield is also an approximate estimate of the volume of water required to saturate an aquifer by flow under influence of gravity.

Specific Retention

Specific retention of a rock or formation is measured as the ratio of volume of water that after saturation is retained against the force of gravity to the total volume of the rock or formation:

$$S_r = \frac{V_r}{V}$$

Where,

- S_r = Specific retention.
- V_r = Volume of water retained against the force of gravity.
- V = Total volume of rock.

A look at previous equation reveals that Specific retention is ratio of two volumes and does not have any unit, rather it is dimensionless.

Here if we add specific yield and specific retention we get equation-5. In this equation, it is clear that V_y: the volume of water yielded plus V_r: the volume of water retained is a measure of total voids or pores in the formation. Hence, it is clear that specific yield added with specific retention equal to porosity of the formation.

$$S_y + S_r = \frac{V_y + V_r}{V} = n(\text{porosity})$$

Storage Coefficient and Storativity

Storage coefficient helps in estimation of the storing capability of an aquifer. Storage coefficient is a general term, which refers to volume of water either taken in or released out by the aquifer per unit surface area per unit change in hydraulic head.

$$S_c = \frac{V_w}{A \times \Delta h}$$

Where,

- S_c refers to storage coefficient.
- V_w refers to the volume of water either taken in or released out by the aquifer.
- A refers to the surface area of the aquifer.
- Δh refers to change in hydraulic head.

Storage coefficients for confined aquifer are also referred to as Storativity.

Specific Storage

Specific storage is defined as the volume of water that an aquifer takes in or releases per unit volume of the aquifer per unit decline in hydraulic head.

$$S_S = \frac{V_w}{V \times \Delta h}$$

Where,

- S_s refers to specific storage.
- V_w refers to the volume of water either taken in or released out by the aquifer.
- V refers to the volume of the aquifer.
- Δh refers to change in hydraulic head

A look at equation $S_c = \frac{V_w}{A \times \Delta h}$ reveals that it has unit of per meter and dimension of L^{-1}.

In case of confined aquifers, the relationship between Storativity and Specific storage is given by:

$$S = S_S \times b$$

Where,

- S is storage coefficient.
- S_s is specific storage.
- b is the saturated thickness of the aquifer.

Hydraulic Conductivity

It refers to aquifer's ability to transmit or conduct water. It is defined as volume rate of water of given kinematic viscosity moving through unit cross sectional area per unit hydraulic gradient. It is to be noted that the unit cross sectional area mentioned above is at right angle to the direction of groundwater flow.

$$K = \frac{Q}{A \times \frac{\Delta h}{\Delta l}}$$

Where,

- K is hydraulic conductivity.
- Q is the volume rate of the water.
- A is the cross sectional area.
- $\frac{\Delta h}{\Delta l}$ is the hydraulic gradient.

It can be observed from previous equation that the unit of hydraulic conductivity is m/day with dimension of LT^{-1}.

It is to be noted that hydraulic conductivity is a function of the porous media and the fluid passing through it.

Intrinsic Permeability

Intrinsic permeability (k) is fundamental property of the aquifer, which determines its ability to transmit any fluid through it. It is a function of media only.

$$k = C \times d^2$$

Where,

- C is a constant dependent on factors like distribution of grain size, sphericity and roundness of grains, nature of their packing etc.
- d is diameter of the grains.

The dimension of intrinsic permeability is L^2 and the popularly used unit is "Darcy", where 1 Darcy $\approx 10^{-8}$ cm^2.

The relationship between hydraulic conductivity and intrinsic permeability can be understood with help of:

$$K = \frac{k \times \rho \times g}{\mu}$$

Where,

- K is hydraulic conductivity.
- k is intrinsic permeability.
- ρ is density.
- g is acceleration due to gravity.
- μ is kinematic viscosity.

Transmissivity

It is yet another property, which refers to aquifer's ability to transmit or conduct water. It is defined as volume rate of water of given kinematic viscosity conducted under influence of unit hydraulic gradient through unit saturated width of the aquifer at right angle to the direction of groundwater flow:

$$T = \frac{Q}{W \times \frac{\Delta h}{\Delta l}}$$

Where,

- T is Transmissivity.
- Q is the volume rate of the water.

- W is the saturated width of the aquifer.
- $\frac{\Delta h}{\Delta l}$ is the hydraulic gradient.

It can be observed from previous equation that the unit of transmissivity is m^2/day with dimension of $L^2 T^{-1}$. As in case of hydraulic conductivity, transmissivity is also a function of the porous media and the fluid passing through it.

Aquifer Characterization Techniques

Electrical Techniques

In geophysical investigations using electrical techniques, two primary properties of interest been considered are electrical conductivity or dielectric constant.

Electrical techniques make involves profiling and sounding mode of data collection which both involves the use of direct currents or low frequency altering currents passing through the subsurface. While profiling use of the terms profiling involves the inferring of subsurface measurements based on lateral changes in electrical properties over constant subsurface, Sounding infers subsurface measurements of single location as a function of changes in petro-physical properties as function of depth. The use of either of the two modes for data collection is function of the purpose of the investigation.

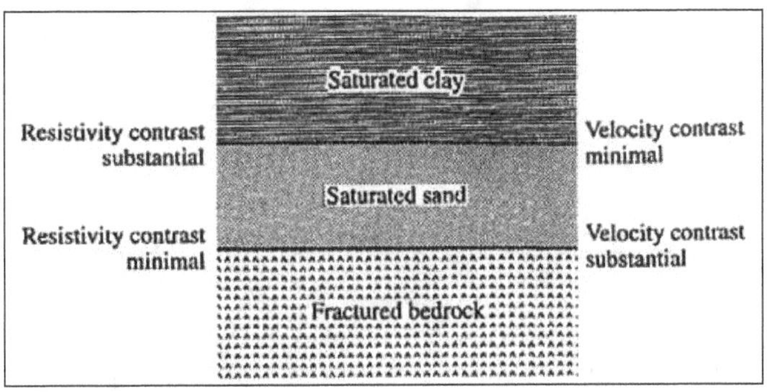

Schematic geological section and associated resistivity and velocity contrasts at interfaces.

Electrical Resistivity

Electrical resistivity (ER) is more frequently been used as compared to other electrical techniques in groundwater investigations of which includes the characterization of aquifers. Electrical Resistivity (ER) involves the introduction of time – varying direct current (DC) or very low frequency (<1 Hz) current into the ground between two current electrodes to generate potential differences as measured at the surface with units of Ohm-meters (Ω-m). A deviation from the norm in the pattern of potential differences expected from homogeneous proffer the necessary information on the form and electrical properties of subsurface inhomogeneities.

A typical Electrical Resistivity (ER) investigation made up of a 2 – electrode system would include 2 – current electrodes and 2 – potential electrodes. As current is been injected into the ground, corresponding potential differences (ΔV) is measured. This measurement coupled with known current (I) and a geometric factor (K) that is a function of the particular electrode configuration, can be used to calculate resistivity (ρ) following Ohm's law:

$$\rho = \frac{\Delta V}{I} K$$

The expression in previous equation for a homogeneous ground is also the same applied for heterogeneous ground; however the general term "apparent resistivity (ρa)" is substituted for resistivity (ρ) in previous equation. Apparent resistivity (ρa) is used here rather than the actual resistivity of the subsurface due to the non-homogeneity nature of the subsurface.

A four – electrode configurations is been used most commonly when it comes to measuring apparent resistivity of the subsurface. The simplest of these configurations is the Wenner configuration where the outer two current electrodes C_1 and C_2, apply a constant current, and the inner two potential electrodes, labeled P_1 and P_2, measures voltage difference created by this current. The electrode spacing has a fixed value a, and the apparent resistivity of the subsurface sampled by this array could be computed using the equation:

$$\rho a = \frac{\Delta V}{I} 2\pi a$$

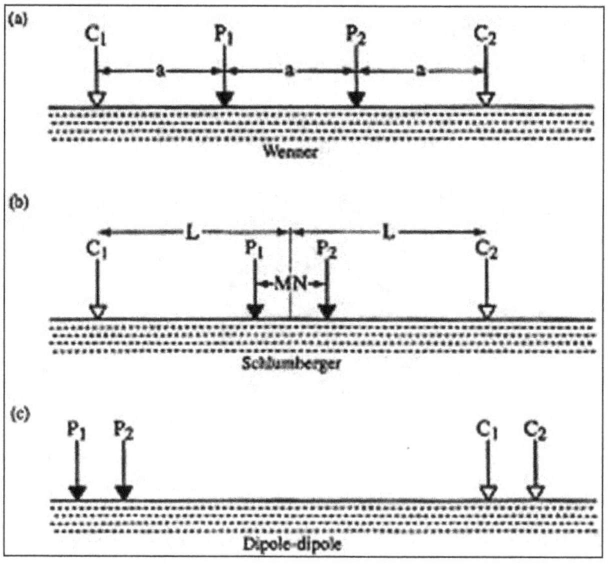

Common electrode configuration used to measure apparent resistivity of the subsurface C_1 and C_2 are the current electrodes and P_1 and P_2 are the potential electrodes. (a) Wenner Array (b) Schlumberger array (c) dipole – Dipole array.

Asides the Wenner array mode of electrode configuration, another commonly used electrode configuration is the Schlumberger array, where the spacing (MN) between the potential electrodes (P_1, P_2) is much smaller as compared to the spacing (2 L) between the current electrodes (C_1, C_2).

The electrode configuration in represents that of a dipole – dipole array where the potential electrode pair and current electrode is closely spaced, however there exist significant distances between the two sets of electrodes. Unlike the cases of the Wenner and Schlumberger arrays, where data collected through either profiling or sounding mode depends a lot on the electrode array geometry.

To illustrate, we consider the Wenner array. Profiling involves the lateral movement of the entire array along the surface at fixed distances to obtain apparent resistivity measurements as a function of distance. The values of the measurements are assigned to the geometric center of the electrode array. Interpretation of measurements is usually with its data aimed at location of geological structures buried stream channels, aquifers or water bearing formations etc.

Sounding unlike profiling involves gradual and progressively expansion of expansion of the array about a fixed central point with current and potential electrodes being maintained at a relative spacing with depth been a function of electrode spacing and subsurface resistivity contrasts. The dashed lines representing current flow lines in an homogeneous environment while bold lines represents actual current flow in single interface that separates units with different resistivities. Next we look how electrode spacing, current and its' influence on depth of penetration. In figure, when the electrode spacing is close, it is observed that the current only upper interface (i.e. the interface of lower resistivity). The scenario in figure is different; as electrode spacing has increased resulting in greater penetration depth and higher apparent resistivity values due to the influence of the lower (higher resistivity) layer. Lastly when the electrodes are farther apart, only substantial current amounts are found to flow through the resistivity layer.

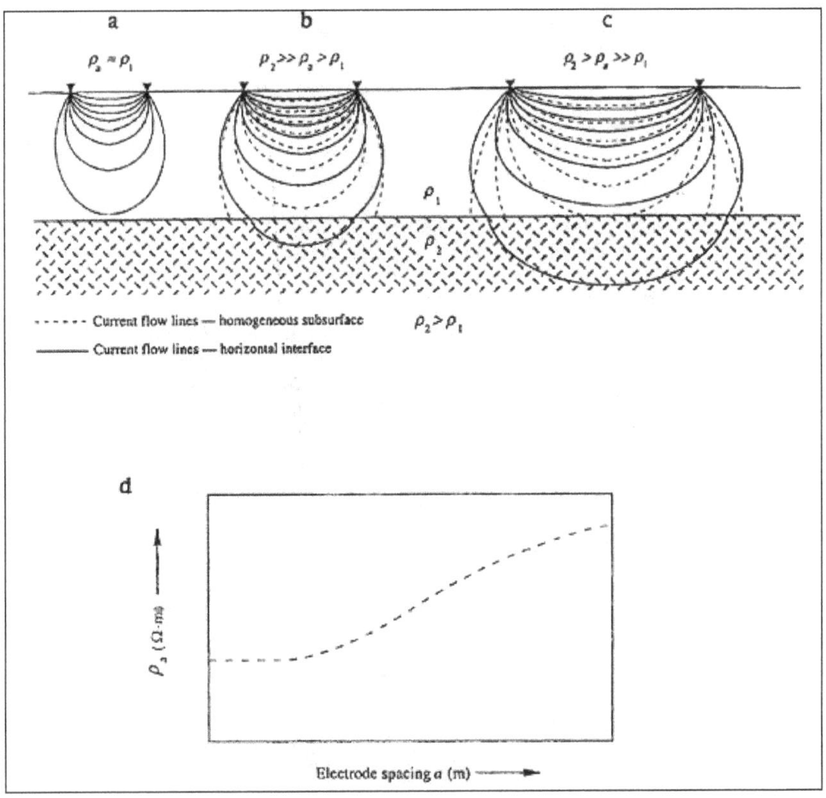

Figure shows effects of electrode spacing and presence of an interface on apparent resistivity measurements. The dashed lines represent current flow lines in the absence of the interface and the solid lines represent actual current flow lines (a–c) As the current electrode spacing is increased, the current lines penetrate deeper and the apparent resistivity measurements are influenced by the lower (more resistive) layer. (d) The qualitative variations in apparent resistivity as a function of electrode spacing are illustrated by the two-layer sounding curve.

The curve reveals qualitative variations in apparent resistivities which increases with electrode spacing, a, this curve is known as a sounding curve revealing geology of the subsurface with resistivity increasing with depth provided geology is homogeneous. However in the case where the geology is inhomogeneous it results in a complex sounding curve whose interpretation is non-unique. To interpret electrical resistivity sounding data, various curve-fitting or computer inversion schemes are used or measured and compared with model computations. A classic example where both modes of data acquisition (profiling and sounding) is been used is in the location of a buried stream channel using Wenner array. The contour map produced from resistivity measurements of several profiles collected near San Jose, CA, using an a-spacing of 6.1 reveals contours of equal apparent resistivity delineating an approximately east–west trending high apparent resistivity values. To understand the cause of high apparent resistivity values here, a geological cross-section (BA) was drawn across the map. The geological cross-section (BA) drawn is based on four expanding – spread traverses (soundings), apparent resistivity profile information and information from three boreholes whose locations are indicated on the cross-section. The critical observation of the cross-section shows that the area with high-resistivity as on the apparent resistivity map is a zone of gravels and boulders that defines the location of a buried stream channel (subsurface structure).

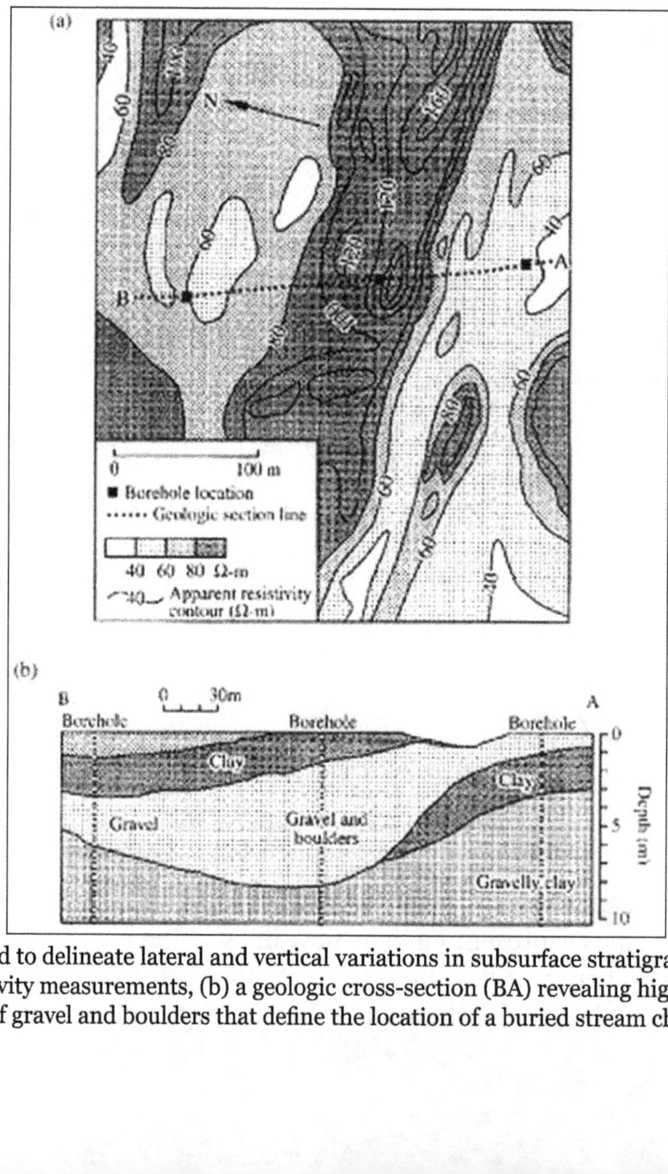

Resistivity survey used to delineate lateral and vertical variations in subsurface stratigraphy. (a) Contour map produced from resistivity measurements, (b) a geologic cross-section (BA) revealing high-resistivity trend in a zone of gravel and boulders that define the location of a buried stream channel.

Aside the mapping of subsurface structure and stratigraphy, electrical resistivity measurements could be channeled towards the inferring lithological information and hydrogeological parameters needed for the mapping groundwater. For groundwater mapping, electrical conduction (inverse of electrical resistivity) is considered. Here the interest is the delineation of connected pore spaces, void spaces, interstices, fractures within rocks that are water filled which leads to a reduced resistivity values and high conductivity. However more information is still needed as high conductivity within rock formation or units could be due to a number of things asides water some of which includes presence of clay minerals, contamination plumes etc.

Common earth materials have wide range of electrical resistivity values revealed in table, however some of these values are known to overlap for different earth materials. Values commonly vary over 12 orders of magnitude and have a maximum range of 24 orders of magnitude. The following statement as regarding electrical resistivity holds:

- Resistivity is sensitive to moisture content; thus unsaturated sediments usually have higher resistivity values than saturated sediments.
- Sandy materials generally have higher resistivity values than clayey materials.
- Granitic bedrock generally has a higher resistivity value than saturated sediments and frequently offers a large apparent resistivity contrast when overlain by these sediments.

Table: Resistivity and dielectric constants for typical near-surface materials.

Material	Resistivity (ohm-m)	Dielectric Constant
Sand (Dry)	10^3-10^7	3-6
Sand (Saturated)	10^2-10^4	20-30
Silts	10^2-10^3	5-30
Shales	10-10^3	5-15
Clays	1-10^3	5-40
Humid Soil	50-100	30
Cultivated Soil	200	15
Rocky Soil	1000	7
Sandy Soil (Dry)	7100	3
Sandy Soil (Saturated)	150	25
Loamy Soil (Dry)	9100	3
Loamy Soil (Saturated)	500	19
Clayey Soil (Dry)	3700	2
Clayey Soil (Saturated)	20	15
Sandstone (Saturated)	25	6
Limestone (Dry)	10^9	7
Limestone (Saturated)	40	4-8
Basalt (Saturated)	100	8
Granite	10^3-105	4-6
Coal	10^4	4-5
Fresh Water	30-10^4	81

Permafrost	10^2-10^5	4-8
Dry Snow	10^5-10^8	1
Ice	10^3-10^5	4-12

Asides the use of sounding curves, empirical formulae have also been adapted in relating measurement of apparent resistivity with hydrological parameters of interest as this relates to aquifers. The empirical formula developed in the laboratory by Archie relates these parameters:

$$\rho_r = a\varnothing^{-m} S^{-n} \rho_w$$

Where, ρ_r is the electrical resistivity of the rock, ρ_w is the pore water resistivity, $\varnothing$ is the fractional porosity, S is the fractional water saturation.

And n, a, and m are constants {$n \approx 2$, $0.6 \leq a \leq 1.0$, and $1.4 \leq m \leq 2.2$; Ward}. Though Archie's law was formulated using lithified materials, Jackson et al. posited its' accurate usability for unconsolidated materials also. The equation presented by previous equation is used generally for well log interpretation however if ρ_r, ρ_w, and $\varnothing$ can be measured separately such that a and m are estimated reasonably then the fractional water saturation could also be inferred using electrical surveys. This concept was utilized by Pfeifer and Anderson to observe and monitor the migration of tracer-spiked water through the subsurface using resistivity array.

In conclusion, it could be said that the complexities that exist in the interpretation of sounding curves and the non-unique solution it gives, suggests the suitability of surface resistivity in determined subsurface geology. Also due to its sensitivity to parameters like moisture content it's been termed a useful tool in hydrological investigations as reviewed by Ward, Van Nostrand and Cook.

Electromagnetic Induction

Electromagnetic (EM) techniques as tool for geophysical exploration has dramatically increased in recent years served as a useful tool for groundwater and environmental site assessment. It involves the propagation of continuous-wave or transient electromagnetic fields in and over the earth through resulting in the generation of time-varying magnetic field. For any of such surveys to be carried out three components are essential; a transmitters, receivers, buried conductors or conductive subsurface. These three form a trio of electric-circuit coupled by an EM induction with currents been introduced into the ground directly or through inductive means by the transmitters.

The Primary field travels from the transmitter coil to the receiver coil via paths above and below the surface. Where a homogenous subsurface is detected no difference is observed between the fields propagated above, below and within the surface other than a slight reduction in amplitude. However, the interaction of the time-varying field with a conductive subsurface induces eddy currents, which gives rise to a secondary magnetic field. The attributes of the fields generated, such as amplitude, orientation and phase shift can be measured by the receiver coil and compared with those of the primary field as such information about the presence of subsurface conductors, or subsurface electrical conductivity distribution can be inferred.

It's paramount to recall that electrical conductivity is an inverse of electrical resistivity; as such electrical conductivity measurements made using electromagnetic methods is also dependent on subsurface texture, porosity, presence of clay minerals, moisture content and the electrical resistivity of the pore fluid presence. The acquisition of EM data requires less time, achieving greater depth of investigation than resistivity techniques. However, the equipment used are expensive and the methods used to qualitatively interpret data from EM surveys is complicated than those used in resistivity methods. This is because a conductive subsurface environment is essential to set up a secondary field measured with inductive EM methods. Electromagnetic methods as a tool for geophysical investigation and exploration is most suited for the detection of water—bearing formation (aquifers) and high – conductive subsurface target such as salt water saturated sediments.

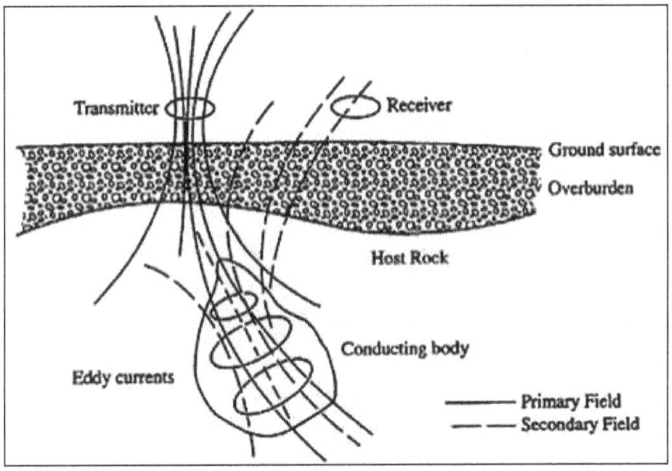

Electromagnetic induction technique.

Instrumentation could take in varying forms; but mainly consist of a source and receiver or receiver units. The source (transmitter) transmits time-varying magnetic fields with the receiver measuring components of the total (primary and secondary) field, magnetic field, sometimes the electric field and the necessary electronic circuitry to process, store and display signals. Data obtained from electromagnetic surveys, like their resistivity counterpart can be collected in profile and sounding mode with their information been presented as maps or pseudo-section to give a better picture of the subsurface. Acquisition, resolution and depth of investigation from this survey are been governed by mostly by conditions of the subsurface and domain of measurement.

EM surveys are divided into two domain system of measurement namely; frequency and time domain system. For frequency domain EM systems, we have the transmitter classed as either high or low frequency transmitters; high transmitter frequencies permits high- resolution investigation of subsurface conductors at near-surface or shallow depths while lower transmitter frequencies allows for deeper depth of investigation at the expense of resolution. This implies that high frequency EM surveys yield better result for near-surface due to high resolution, however if interested in deeper subsurface investigation (low frequency EM surveys) then we have need a way around the low resolution. In the case of time domain system, secondary magnetic field is measured as a function of time, with early – time measurement being suited best for near-surface information while late- time measurement yields results of the deeper subsurface. It is paramount to note that depth of penetration or investigation and resolution is also been governed by coil configuration; while measurements from coil separations are influenced by electrical properties thus the larger coil separation investigates greater depths while smaller coil separation investigates near-surface.

Because Electrical Conductivity is related inversely to Electrical Resistivity, as such discussions relating electrical resistivity to lithology or hydrological properties can be applied in an inverse manner to measurements involving electrical conductivity. Electrical conductivity for example is higher for saturated sediments, clayey materials than for unsaturated sediments and sandy materials respectively. Some examples of investigations involving EM surveys include Sheets and Hendricks, who used EM induction methods to estimate soil water content and McNeill that discussed the relation between electrical conductivity and hydrogeological parameters of porosity and saturation.

Ground Waves Techniques

Ground Penetrating Radar

Ground Penetrating Radar (GPR) as a geophysical technique is relative new and becoming increasingly popular critically understanding the events of the near-surface or shallow subsurface. Davis and Annan viewed the Ground Penetrating Radar (GPR) as a technique of imaging the subsurface at high resolution using electromagnetic waves transmitted at frequencies between 10 to 1000 MHz. GPR could also be viewed as a non-destructive geophysical technique due to its successful geological applications in urban and sensitive environments. Some of these applications include the subsurface mapping of water table soils and rocks structures (e.g. groundwater channels) at high resolutions. It is similar in principle to seismic reflection profiling in however, propagation of radar waves through the subsurface is controlled by electrical properties at high frequencies.

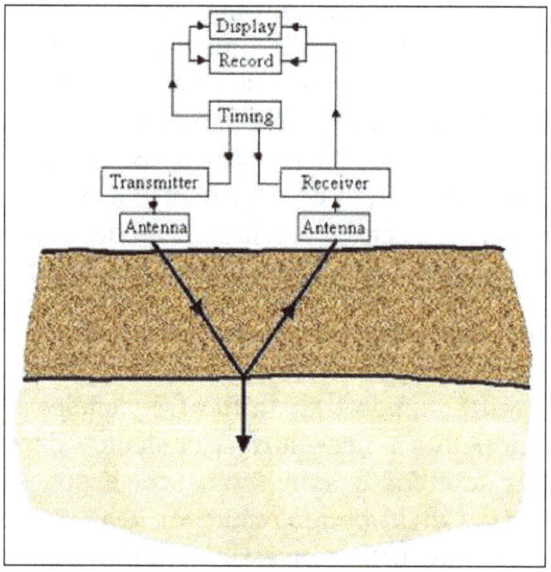

Flow chart for a typical GPR system.

The GPR survey system is made up of three vital components; a transmitter, a receiver directly connected to the antenna and the control unit. The transmitter radiates EM waves into the subsurface that could be refracted, diffracted or primarily reflected depending on the dielectric permittivity and electrical conductivity nature of the subsurface interfaces encountered. Recorded radar data received after the survey is first been observed, analyzed and interpreted by the aid of inbuilt radar processing software like RADPro, Ekko depending on the system type and make. These data are presented in form of radargrams which could either be presented as 2D or 3D subsurface images depending on the combination of the different axes (x, y and z) involved. Interpreting of radargrams is performed by interface mapping which is quite similar to the technique used in interpreting of seismograms. Here each band within on a radargram is presumably classed and identified as a distinct geological horizon; this would have been correct except for the effects of multiples, interference with previous reflections, noise etc. All this effects on the radargram need to be removed to correctly identify the different geological horizons and geological structures as present within the radargram as such radargram are subjected to varying radar processing operations depending on the aims, objective of the survey been undertaking through the help of inbuilt system radar processing software like RADpro, Pulse Ekko system software etc.

Processing of the radargram could be simplified by processing operations such as dewowing (removal of low frequency components), Gain Control (strengthen weaker events), deconvolution (restores shape of downgoing wave train such that primary events could be recognized more easily), Migration (useful in removing diffraction hyperbolae and restoring dips). The resultant radargram when correlated with the subsurface geology shows varying interfaces, geological structures that might be present. Though GPR has successfully been utilized in unsaturated (non-electrically conductive or highly resistive) and saturated (electrically conductive) environment, however performance is higher in unsaturated (non-conductive) than in saturated (conductive) such as non-expanding clay environment such as at Savannah River Site in South Carolina.

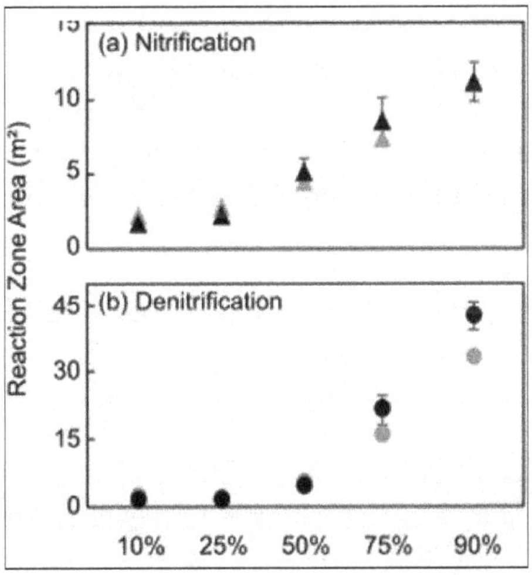

(A) Interpretation of a GPR profile image (B) Interpretation of the prominent stratigraphic units, structures and faults.

The depth of penetration or investigation of GPR survey is function of the frequency of the EM waves or radar waves and nature of the subsurface material been investigated as shown in figure for varying subsurface materials at frequencies ranging between 1 and 500 MHz. If the nature of subsurface material is highly resistive and has low conductivity then we expect a higher depth of penetration however for subsurface materials that are less resistivity and very conductive we expect low depth of penetration. Depth of penetration asides from been dependent on nature of the subsurface material (i.e. resistivity or conductivity nature) is also a function of frequencies which in turn affects resolution of subsurface imagery or radargram. Thus at low frequencies, we expect a greater depth of penetration at the expense of resolution while at high frequencies, we achieve a lower depth of penetration at higher resolution.

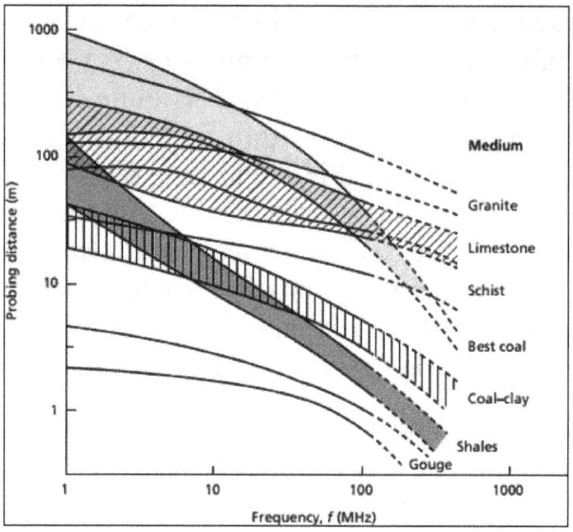

The relationship between probing distance and frequency for different materials.

Ground Penetrating Radar (GPR) data have been successful utilized in the hydrogeological investigations to locate the water table and to delineate shallow, unconsolidated aquifers.

Seismic Techniques

The use of Seismic techniques in subsurface characterization is based on the propagation of elastic waves generated from a seismic controlled source, propagated through the subsurface, boreholes, received by receivers (geophones or hydrophones) and displayed on seismographs (as a combination of waves velocities and attenuation). From these, properties of the subsurface like porosity, hydraulic conductivity, elastic moduli and water saturation which could help us better understand the subsurface could be derived.

Subsurface investigation involving Seismic techniques are categorized into three; Seismic Refraction, Cross-hole transmission (tomography) and Seismic Reflection.

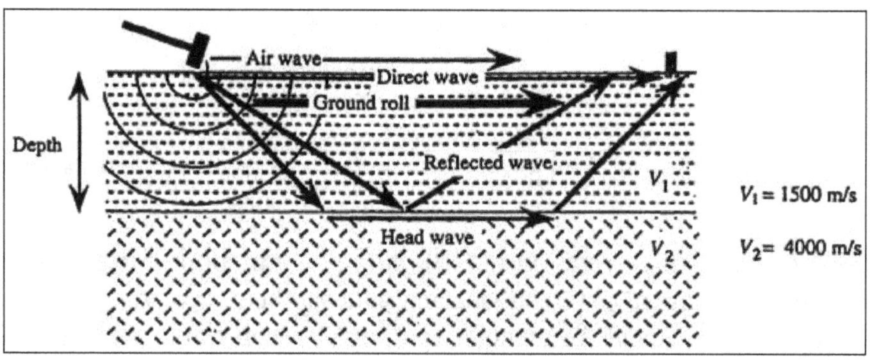

Major ray paths of P-wave energy.

With Seismic Refraction, the incident ray is refracted along the target boundary before returning to the surface. The arrival times gotten from the refracted energy are displayed as function of distance from the source with their interpretation been made manually using simple software or forward modeling techniques. The relationship between arrival times and distances could be used to obtain velocity information directly. Seismic Refraction techniques are the most appropriate for a few shallow (50 m) targets of interest, or where one is interested in identifying gross lateral velocity variations or changes in interface dip. Though Seismic Refraction yields lower resolution than Seismic Reflection and Seismic Cross-hole tomographic, it is however chosen over Reflection as they are inexpensive and help to determining the depth to the water table (buried refractor) and to the top of bedrock, the gross velocity structure, or for locating significant faults. The buried refractor is usually saturated and has a greater velocity than the unsaturated equivalent soil unit and the bedrock surface.

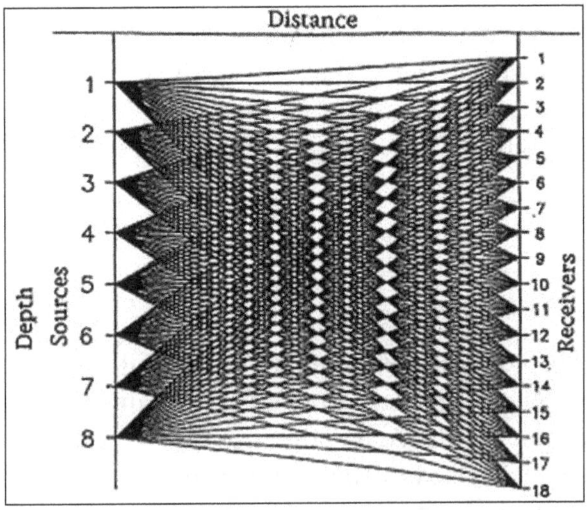

Cross-hole tomography geometry for seismic and radar methods. Sources and receivers are located in separated boreholes, and energy from each source is received by all geophones. Cross-holes acquisition geometries have also been used with electrical resistivity and EM methods.

Cross-hole transmission (tomography) data acquisition is possible using several techniques amongst which includes seismic techniques, electrical resistivity, electromagnetic, radar with seismic being the most common. Majority of cross – hole tomographic seismic data have been collected for research however the those collected over extremely high resolution of up to 0.5 m are better suited for site characterization. Figure shows a typical example of seismic cross-hole survey. The multiple sampling of the intra-wellbore area permits very detailed estimation of the velocity structure. As seismic P-wave velocities can be related to lithological and hydrogeological parameters, this extremely high resolution method is ideal for detailed stratigraphic and hydraulic characterization of interwell areas.

References

- Aquifer-classification-and-characterization, aquifers-matrix-and-fluids: intechopen.com, Retrieved 03, March 2020
- Aquifer-can-be-basically-classified-focus-of-any-groundwater-exploration: iasri.res.in, Retrieved 18, July 2020

4

Groundwater and Surface Water

The water present under the surface of the Earth in the cracks and spaces of soil is known as groundwater. The water located on the surface of Earth in the form of streams, rivers, lakes, reservoirs, wetlands, etc. is known as surface water. The chapter closely examines the key concepts of groundwater and surface water to provide an extensive understanding of the subject.

Groundwater

The groundwater all over the world has emerged as a potential source for domestic and irrigation purposes. In this case, it becomes important to assess the quality of groundwater before assigning it for domestic or irrigation purposes. We must realize that the natural groundwater without any anthropogenic influences and will also have a baseline concentrating of different dissolved solutes. These solutes are primarily introduced in to the groundwater system by the process of sediment groundwater interaction. This natural groundwater quality varies from place to place depending upon nature of soil, nature of subsurface lithology, climatic conditions etc. The dissolved solids in groundwater can be divided in major ions, minor constituents and trace elements. In general, the major ions of groundwater include sodium, potassium, calcium, magnesium cations and bicarbonate, carbonate, chloride and sulphate anions. The minor constituents include small amount of dissolved solids like iron, manganese, fluoride etc. Besides many elements of the periodic table are present in very small concentration in the groundwater and they are referred to as trace elements. Further, at some places we can also find organic compound in groundwater.

Groundwater represents about 30% of world's fresh water. From the other 70%, nearly 69% is captured in the ice caps and mountain snow/glaciers and merely 1% is found in river and lakes. Groundwater counts in average for one third of the fresh water consumed by humans, but at some parts of the world, this percentage can reach up to 100%. In the illustration bellow an overview is given of Earth's water distribution.

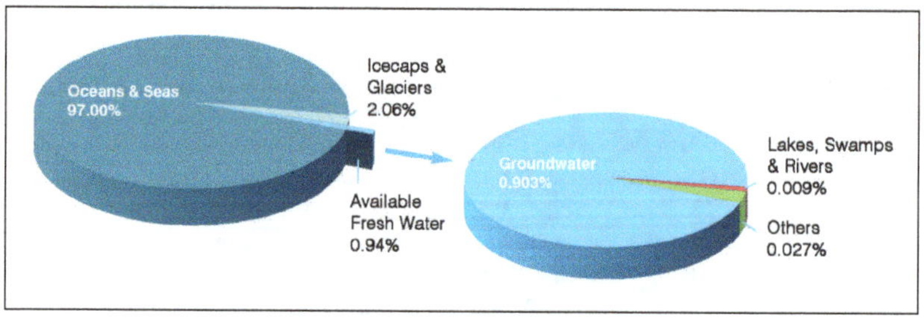

Earth's water distribution.

Groundwater is a very important natural resource and has a significant role in the economy. It is the main source of water for irrigation and the food industry. In general groundwater is a reliable source of water for the agriculture and can be used in a flexible manner: when it's dry and there is larger demand more groundwater can be extracted and when the rain fall meets the necessities, less groundwater will need to be extracted. Globally, irrigation accounts for more than 70% of total water withdraw (both surface and groundwater). Groundwater is estimated to be used for circa 43% of the total irrigation water use.

For the environment groundwater plays a very important role in keeping the water level and flow into rivers, lakes and wetlands. Specially during the drier months when there is little direct recharge from rainfall, it provides the environment with groundwater flow through the bottom of these water bodies and becomes essential for the wild life and plants living in these environment. Groundwater also plays a very relevant role in sustain navigation through inland waters in the drier seasons. By discharging groundwater into the rivers it helps keeping the water levels higher.

Groundwater is found almost everywhere and its quality is usually very good. The fact that groundwater is stored in the layers beneath the surface, and sometime at very high depths, helps protecting it from contamination and preserve its quality. Additionally, groundwater is a natural resource which can often be found close to the final consumers and therefore does not require large investments in terms of infrastructure and treatment, as it often is necessary when harvesting surface water. The most important about using groundwater is to find the right balance between withdrawing and letting the aquifer's level recover to avoid overexploitation and to avoid pollution of this crucial resource.

Parameters of Water Quality

There are three types of water quality parameters physical, chemical, and biological.

Physical Parameters of Water Quality

Turbidity

Turbidity is the cloudiness of water. It is a measure of the ability of light to pass through water. It is caused by suspended material such as clay, silt, organic material, plankton, and other particulate materials in water.

Turbidity in drinking water is esthetically unacceptable, which makes the water look unappetizing. The impact of turbidity can be summarized in the following points:

- It can increase the cost of water treatment for various uses.
- The particulates can provide hiding places for harmful microorganisms and thereby shield them from the disinfection process.
- Suspended materials can clog or damage fish gills, decreasing its resistance to diseases, reducing its growth rates, affecting egg and larval maturing, and affecting the efficiency of fish catching method.
- Suspended particles provide adsorption media for heavy metals such as mercury, chromium, lead, cadmium, and many hazardous organic pollutants such as polychlorinated biphenyls (PCBs), polycyclic aromatic hydrocarbons (PAHs), and many pesticides.
- The amount of available food is reduced because higher turbidity raises water temperatures

in light of the fact that suspended particles absorb more sun heat. Consequently, the concentration of the dissolved oxygen (DO) can be decreased since warm water carries less dissolved oxygen than cold water.

- Turbidity is measured by an instrument called nephelometric turbidimeter, which expresses turbidity in terms of NTU or TU. A TU is equivalent to 1 mg/L of silica in suspension.
- Turbidity more than 5 NTU can be visible to the average person while turbidity in muddy water, it exceeds 100 NTU. Groundwater normally has very low turbidity because of the natural filtration that occurs as the water penetrates through the soil.

Temperature

Palatability, viscosity, solubility, odors, and chemical reactions are influenced by temperature. Thereby, the sedimentation and chlorination processes and biological oxygen demand (BOD) are temperature dependent. It also affects the biosorption process of the dissolved heavy metals in water. Most people find water at temperatures of 10–15 °C most palatable.

Color

Materials decayed from organic matter, namely, vegetation and inorganic matter such as soil, stones, and rocks impart color to water, which is objectionable for esthetic reasons, not for health reasons.

Color is measured by comparing the water sample with standard color solutions or colored glass disks. One color unit is equivalent to the color produced by a 1 mg/L solution of platinum (potassium chloroplatinate (K_2PtCl_6)). The color of a water sample can be reported as follows:

- Apparent color is the entire water sample color and consists of both dissolved and suspended components color.
- True color is measured after filtering the water sample to remove all suspended material.

Color is graded on scale of 0 (clear) to 70 color units. Pure water is colorless, which is equivalent to 0 color units.

Taste and Odor

Taste and odor in water can be caused by foreign matter such as organic materials, inorganic compounds, or dissolved gasses. These materials may come from natural, domestic, or agricultural sources.

The numerical value of odor or taste is determined quantitatively by measuring a volume of sample A and diluting it with a volume of sample B of an odor-free distilled water so that the odor of the resulting mixture is just detectable at a total mixture volume of 200 ml. The unit of odor or taste is expressed in terms of a threshold number as follows:

$$\text{TON or TTN} = (A + B)/A$$

Where, TON is the threshold odor number and TTN is the threshold taste number.

Solids

Solids occur in water either in solution or in suspension. These two types of solids can be identified by using a glass fiber filter that the water sample passes through. By definition, the suspended solids are retained on the top of the filter and the dissolved solids pass through the filter with the water.

If the filtered portion of the water sample is placed in a small dish and then evaporated, the solids as a residue. This material is usually called total dissolved solids or TDS.

$$\text{Total solid}\left(\text{TS}\right) = \text{Total dissolved solid}\left(\text{TDS}\right) + \text{Total suspended solid}\left(\text{TSS}\right)$$

Water can be classified by the amount of TDS per liter as follows:

- Freshwater: <1500 mg/L TDS;
- Brackish water: 1500–5000 mg/L TDS;
- Saline water: >5000 mg/L TDS.

The residue of TSS and TDS after heating to dryness for a defined period of time and at a specific temperature is defined as fixed solids. Volatile solids are those solids lost on ignition (heating to 550 °C).

These measures are helpful to the operators of the wastewater treatment plant because they roughly approximate the amount of organic matter existing in the total solids of wastewater, activated sludge, and industrial wastes. Figure describes the interrelationship of solids found in water. They are calculated as follows:

1. Total solids:

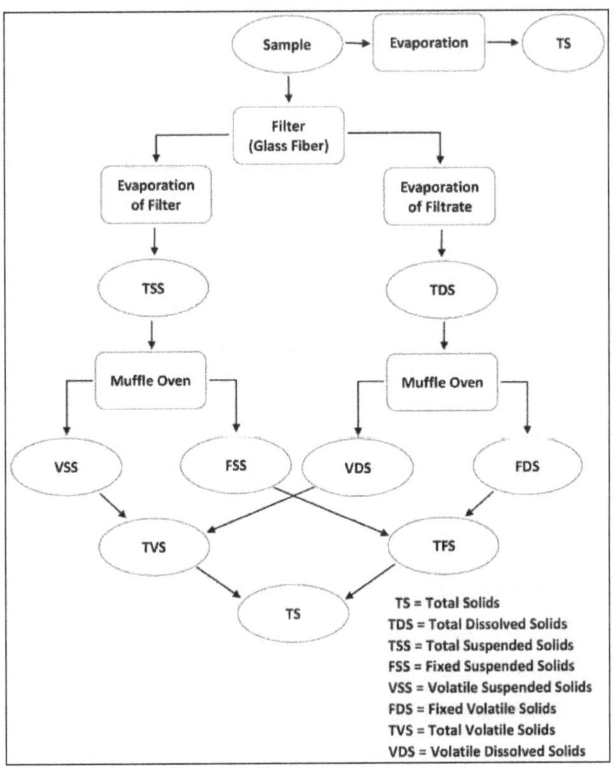

Interrelationship of solids found in water.

$$\text{Total solids}(\text{mg}/\text{L}) = [(\text{TSA} - \text{TSB})] \times 1000/\text{sample}(\text{mL})$$

Where, TSA = weight of dried residue + dish in milligrams and TSB = weight of dish in milligrams.

2. Total dissolved solids:

$$\text{Total dissolved solids}(\text{mg}/\text{L}) = [(\text{TDSA} - \text{TDSB})] \times 1000/\text{sample}(\text{mL})$$

Where, TDSA = weight of dried residue + dish in milligrams and TDSB = weight of dish in milligrams.

3. Total suspended solids:

$$\text{Total suspended solids}(\text{mg}/\text{L}) = [(\text{TSSA} - \text{TSSB})] \times 1000/\text{sample}(\text{mL})$$

Where, TSSA = weight of dish and filter paper + dried residue and TSSB = weight of dish and filter paper in milligram.

4. Fixed and volatile suspended solids:

$$\text{Volatile suspended solids}(\text{mg}/\text{L}) = [(\text{VSSA} - \text{VSSB})] \times 1000/\text{sample}(\text{mL})$$

Where, VSSA = weight of residue + dish and filter before ignition, mg and VSSB = weight of residue + dish and filter after ignition, mg.

Electrical Conductivity (EC)

The electrical conductivity (EC) of water is a measure of the ability of a solution to carry or conduct an electrical current. Since the electrical current is carried by ions in solution, the conductivity increases as the concentration of ions increases. Therefore, it is one of the main parameters used to determine the suitability of water for irrigation and firefighting.

Units of its measurement are as follows:

- U.S. units = micromhos/cm;
- S.I. units = milliSiemens/m (mS/m) or dS/m (deciSiemens/m).

Where, (mS/m) = 10 umho/cm (1000 μS/cm = 1 dS/m).

Pure water is not a good conductor of electricity. Typical conductivity of water is as follows:

- Ultra-pure water: 5.5 × 10−6 S/m;
- Drinking water: 0.005–0.05 S/m;
- Seawater: 5 S/m.

The electrical conductivity can be used to estimate the TDS value of water as follows:

TDS(mg/L) ≅ EC(dS/mor umho/cm) × (0.55 – 0.7)E7

TDS can be used to estimate the ionic strength of water in the applications of groundwater re-charging by treated wastewater. The normal method of measurement is electrometric method.

Chemical Parameters of Water Quality

pH

pH is one of the most important parameters of water quality. It is defined as the negative logarithm of the hydrogen ion concentration. It is a dimensionless number indicating the strength of an acidic or a basic solution. Actually, pH of water is a measure of how acidic/basic water is. Acidic water contains extra hydrogen ions (H^+) and basic water contains extra hydroxyl (OH^-) ions.

As shown in figure, pH ranges from 0 to 14, with 7 being neutral. pH of less than 7 indicates acidity, whereas a pH of greater than 7 indicates a base solution. Pure water is neutral, with a pH close to 7.0 at 25 °C. Normal rainfall has a pH of approximately 5.6 (slightly acidic) owing to atmospheric carbon dioxide gas. Safe ranges of pH for drinking water are from 6.5 to 8.5 for domestic use and living organisms need.

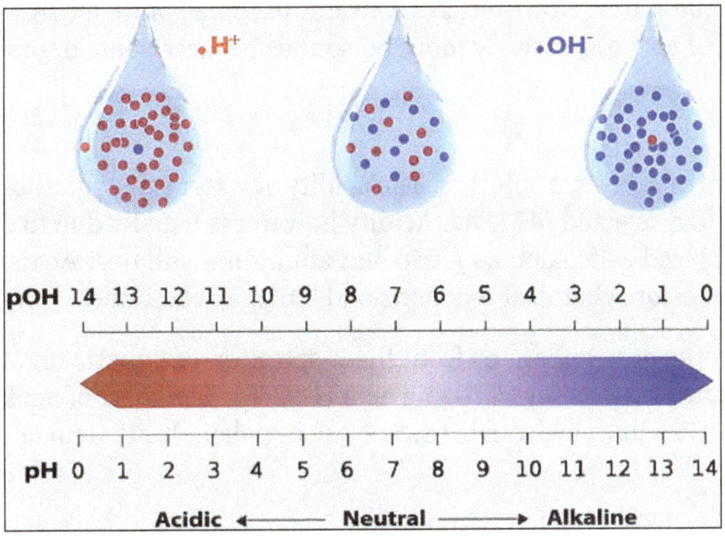

pH of water.

A change of 1 unit on a pH scale represents a 10-fold change in the pH, so that water with pH of 7 is 10 times more acidic than water with a pH of 8, and water with a pH of 5 is 100 times more acidic than water with a pH of 7. There are two methods available for the determination of pH: electrometric and colorimetric methods.

Excessively high and low pHs can be detrimental for the use of water. A high pH makes the taste bitter and decreases the effectiveness of the chlorine disinfection, thereby causing the need for additional chlorine. The amount of oxygen in water increases as pH rises. Low-pH water will corrode or dissolve metals and other substances. Pollution can modify the pH of water, which can damage animals and plants that live in the water. The effects of pH on animals and plants can be summarized as follows:

- Most aquatic animals and plants have adapted to life in water with a specific pH and may suffer from even a slight change.

- Even moderately acidic water (low pH) can decrease the number of hatched fish eggs, irritate fish and aquatic insect gills, and damage membranes.
- Water with very low or high pH is fatal. A pH below 4 or above 10 will kill most fish, and very few animals can endure water with a pH below 3 or above 11.
- Amphibians are extremely endangered by low pH because their skin is very sensitive to contaminants. Some scientists believe that the current decrease in amphibian population throughout the globe may be due to low pH levels induced by acid rain.

The effects of pH on other chemicals in water can be summarized as follows:

- Heavy metals such as cadmium, lead, and chromium dissolve more easily in highly acidic water (lower pH). This is important because many heavy metals become much more toxic when dissolved in water.
- A change in the pH can change the forms of some chemicals in the water. Therefore, it may affect aquatic plants and animals. For instance, ammonia is relatively harmless to fish in neutral or acidic water. However, as the water becomes more alkaline (the pH increases), ammonia becomes progressively more poisonous to these same organisms.

Acidity

Acidity is the measure of acids in a solution. The acidity of water is its quantitative capacity to neutralize a strong base to a selected pH level. Acidity in water is usually due to carbon dioxide, mineral acids, and hydrolyzed salts such as ferric and aluminum sulfates. Acids can influence many processes such as corrosion, chemical reactions and biological activities.

Carbon dioxide from the atmosphere or from the respiration of aquatic organisms causes acidity when dissolved in water by forming carbonic acid (H_2CO_3). The level of acidity is determined by titration with standard sodium hydroxide (0.02 N) using phenolphthalein as an indicator.

Alkalinity

The alkalinity of water is its acid-neutralizing capacity comprised of the total of all titratable bases. The measurement of alkalinity of water is necessary to determine the amount of lime and soda needed for water softening (e.g., for corrosion control in conditioning the boiler feed water). Alkalinity of water is mainly caused by the presence of hydroxide ions (OH^-), bicarbonate ions (HCO^{3-}), and carbonate ions (CO_3^{2-}), or a mixture of two of these ions in water. As stated in the following equation, the possibility of OH^- and HCO_3^- ions together is not possible because they react together to produce CO_3^{2-} ions:

$$OH^- + HCO_3^- \rightarrow CO_3^{2-} + H_2O$$

Alkalinity is determined by titration with a standard acid solution (H_2SO_4 of 0.02 N) using selective indicators (methyl orange or phenolphthalein).

The high levels of either acidity or alkalinity in water may be an indication of industrial or chemical pollution. Alkalinity or acidity can also occur from natural sources such as volcanoes. The acidity and alkalinity in natural waters provide a buffering action that protects fish and other aquatic organisms from sudden changes in pH. For instance, if an acidic chemical has somehow contaminated a lake that had natural alkalinity, a neutralization reaction occurs between the acid and alkaline substances; the pH of the lake water remains unchanged. For the protection of aquatic life, the buffering capacity should be at least 20 mg/L as calcium carbonate.

Chloride

Chloride occurs naturally in groundwater, streams, and lakes, but the presence of relatively high chloride concentration in freshwater (about 250 mg/L or more) may indicate wastewater pollution. Chlorides may enter surface water from several sources including chloride-containing rock, agricultural runoff, and wastewater.

Chloride ions Cl^- in drinking water do not cause any harmful effects on public health, but high concentrations can cause an unpleasant salty taste for most people. Chlorides are not usually harmful to people; however, the sodium part of table salt has been connected to kidney and heart diseases. Small amounts of chlorides are essential for ordinary cell functions in animal and plant life.

Sodium chloride may impart a salty taste at 250 mg/L; however, magnesium or calcium chloride are generally not detected by taste until reaching levels of 1000 mg/L. Standards for public drinking water require chloride levels that do not exceed 250 mg/L. There are many methods to measure the chloride concentration in water, but the normal one is the titration method by silver nitrate.

Chlorine Residual

Chlorine (Cl_2) does not occur naturally in water but is added to water and wastewater for disinfection. While chlorine itself is a toxic gas, in dilute aqueous solution, it is not harmful to human health. In drinking water, a residual of about 0.2 mg/L is optimal. The residual concentration which is maintained in the water distribution system ensures good sanitary quality of water.

Chlorine can react with organics in water forming toxic compounds called trihalomethanes or THMs, which are carcinogens such as chloroform $CHCl_3$. Chlorine residual is normally measured by a color comparator test kit or spectrophotometer.

Sulfate

Sulfate ions (SO_4^{2-}) occur in natural water and in wastewater. The high concentration of sulfate in natural water is usually caused by leaching of natural deposits of sodium sulfate (Glauber's salt) or magnesium sulfate (Epson salt). If high concentrations are consumed in drinking water, there may be objectionable tastes or unwanted laxative effects, but there is no significant danger to public health.

Nitrogen

There are four forms of nitrogen in water and wastewater: organic nitrogen, ammonia nitrogen, nitrite nitrogen, and nitrate nitrogen. If water is contaminated with sewage, most of the nitrogen is in the forms of organic and ammonia, which are transformed by microbes to form nitrites and nitrates. Nitrogen in the nitrate form is a basic nutrient to the growth of plants and can be a growth-limiting nutrient factor.

A high concentration of nitrate in surface water can stimulate the rapid growth of the algae which degrades the water quality. Nitrates can enter the groundwater from chemical fertilizers used in the agricultural areas. Excessive nitrate concentration (more than 10 mg/L) in drinking water causes an immediate and severe health threat to infants. The nitrate ions react with blood hemoglobin, thereby reducing the blood's ability to hold oxygen which leads to a disease called blue baby or methemoglobinemia.

Fluoride

A moderate amount of fluoride ions (F^-) in drinking water contributes to good dental health. About 1.0 mg/L is effective in preventing tooth decay, particularly in children.

Excessive amounts of fluoride cause discolored teeth, a condition known as dental fluorosis. The maximum allowable levels of fluoride in public water supplies depend on local climate. In the warmer regions of the country, the maximum allowable concentration of fluoride for potable water is 1.4 mg/L; in colder climates, up to 2.4 mg/L is allowed. There are four methods to determine ion fluoride in water; the selection of the used method depends on the type of water sample.

Iron and Manganese

Although iron (Fe) and manganese (Mn) do not cause health problems, they impart a noticeable bitter taste to drinking water even at very low concentration. These metals usually occur in groundwater in solution as ferrous (Fe^{2+}) and manganous (Mn^{2+}) ions. When these ions are exposed to air, they form the insoluble ferric (Fe^{3+}) and manganic (Mn^{3+}) forms making the water turbid and unacceptable to most people.

These ions can also cause black or brown stains on laundry and plumbing fixtures. They are measured by many instrumental methods such as atomic absorption spectrometry, flame atomic absorption spectrometry, cold vapor atomic absorption spectrometry, electrothermal atomic absorption spectrometry, and inductively coupled plasma (ICP).

Copper and Zinc

Copper (Cu) and zinc (Zn) are nontoxic if found in small concentrations. Actually, they are both essential and beneficial for human health and growth of plants and animals. They can cause undesirable tastes in drinking water. At high concentrations, zinc imparts a milky appearance to the water. They are measured by the same methods used for iron and manganese measurements.

Hardness

Hardness is a term used to express the properties of highly mineralized waters. The dissolved minerals in water cause problems such as scale deposits in hot water pipes and difficulty in producing lather with soap.

Calcium (Ca^{2+}) and magnesium (Mg^{2+}) ions cause the greatest portion of hardness in naturally occurring waters. They enter water mainly from contact with soil and rock, particularly limestone deposits. These ions are present as bicarbonates, sulfates, and sometimes as chlorides and nitrates. Generally, groundwater is harder than surface water. There are two types of hardness:

- Temporary hardness which is due to carbonates and bicarbonates can be removed by boiling.
- Permanent hardness which is remaining after boiling is caused mainly by sulfates and chlorides.

Water with more than 300 mg/L of hardness is generally considered to be hard, and more than 150 mg/L of hardness is noticed by most people, and water with less than 75 mg/L is considered to be soft.

From health viewpoint, hardness up to 500 mg/L is safe, but more than that may cause a laxative effect. Hardness is normally determined by titration with ethylene diamine tetra acidic acid or (EDTA) and Eriochrome Black and Blue indicators. It is usually expressed in terms of mg/L of $CaCO_3$.

$$\text{Total hardness mg / Las } CaCO_3 = \text{Calcium hardness mg / Las } CaCO_3 + \text{Magnesium hardness mg / Las } CaCO_3$$

An accepted water classification according to its hardness is as in Table.

Table: Parameters of water quality.

No.	Types of water quality parameters		
	Physical parameters	Chemical parameters	Biological parameters
1	Turbidity	pH	Bacteria
2	Temperature	Acidity	Algae
3	Color	Alkalinity	Viruses
4	Taste and odor	Chloride	Protozoa
5	Solids	Chlorine residual	
6	Electrical conductivity (EC)	Sulfate	
7		Nitrogen	
8		Fluoride	
9		Iron and manganese	
10		Copper and zinc	
11		Hardness	
12		Dissolved oxygen	
13		Biochemical oxygen demand (BOD)	
14		Chemical oxygen demand (COD)	
15		Toxic inorganic substances	
16		Toxic organic substances	
17		Radioactive substances	

Table: Classification of water according to its hardness.

Water classification	Total hardness concentration as mg/L as $CaCO_3$
Soft water	<50 mg/L as $CaCO_3$
Moderately hard	50–150 mg/L as $CaCO_3$
Hard water	150–300 mg/L as $CaCO_3$
Very hard	>300 mg/L as $CaCO_3$

Dissolved Oxygen

Dissolved oxygen (DO) is considered to be one of the most important parameters of water quality in streams, rivers, and lakes. It is a key test of water pollution. The higher the concentration of dissolved oxygen, the better the water quality. Oxygen is slightly soluble in water and very sensitive to temperature. For example, the saturation concentration at 20 °C is about 9 mg/L and at 0 °C is 14.6 mg/L. The actual amount of dissolved oxygen varies depending on pressure, temperature, and salinity of the water. Dissolved oxygen has no direct effect on public health, but drinking water with very little or no oxygen tastes unpalatable to some people.

There are three main methods used for measuring dissolved oxygen concentrations: the colorimetric method—quick and inexpensive, the Winkler titration method—traditional method, and the electrometric method.

Biochemical Oxygen Demand (BOD)

Bacteria and other microorganisms use organic substances for food. As they metabolize organic material, they consume oxygen. The organics are broken down into simpler compounds, such as CO_2 and H_2O, and the microbes use the energy released for growth and reproduction.

When this process occurs in water, the oxygen consumed is the DO in the water. If oxygen is not continuously replaced by natural or artificial means in the water, the DO concentration will reduce as the microbes decompose the organic materials. This need for oxygen is called the biochemical oxygen demand (BOD). The more organic material there is in the water, the higher the BOD used by the microbes will be. BOD is used as a measure of the power of sewage; strong sewage has a high BOD and weak sewage has low BOD.

The complete decomposition of organic material by microorganisms takes time, usually 20 d or more under ordinary circumstances. The quantity of oxygen used in a specified volume of water to fully decompose or stabilize all biodegradable organic substances is called the ultimate BOD or BOD_L.

BOD is a function of time. At time = 0, no oxygen will have been consumed and the BOD = 0. As each day goes by, oxygen is used by the microbes and the BOD increases. Ultimately, the BODL is reached and the organic materials are completely decomposed.

A graph of the BOD versus time is illustrated as in figure. This is called the BOD curve, which can be expressed mathematically by the following equation:

$$BOD_t = BOD_L \times \left(1 - 10^{-kt}\right)$$

Where, BOD_t = BOD at any time t, mg/L; BOD_L = ultimate BOD, mg/L; k = a constant representing the rate of the BOD reaction; t = time, d.

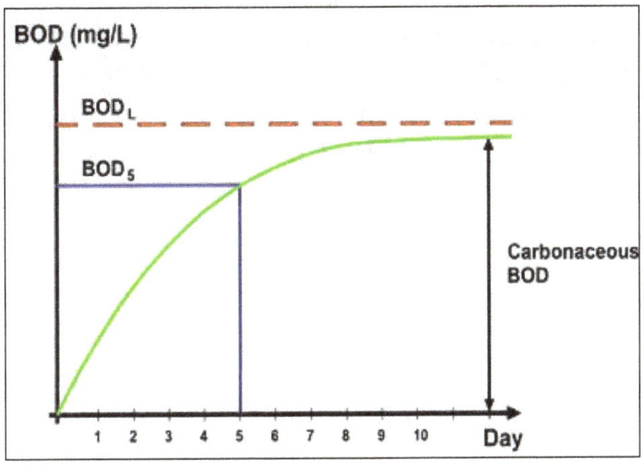

BOD curve

The value of the constant rate k depends on the temperature, the type of organic materials, and the type of microbes exerting the BOD.

Chemical Oxygen Demand (COD)

The chemical oxygen demand (COD) is a parameter that measures all organics: the biodegradable and the non-biodegradable substances. It is a chemical test using strong oxidizing chemicals (potassium dichromate), sulfuric acid, and heat, and the result can be available in just 2 h. COD values are always higher than BOD values for the same sample.

Toxic Inorganic Substances

A wide variety of inorganic toxic substances may be found in water in very small or trace amounts. Even in trace amounts, they can be a danger to public health. Some toxic substances occur from natural sources but many others occur due to industrial activities and/or improper management of hazardous waste. They can be divided into two groups:

- Metallic compounds: This group includes some heavy metals that are toxic, namely, cadmium (Cd), chromium (Cr), lead (Pb), mercury (Hg), silver (Ag), arsenic (As), barium (Ba), thallium (Tl), and selenium (Se). They have a wide range of dangerous effects that differ from one metal to another. They may be acute fatal poisons such as (As) and (Cr^{6+}) or may produce chronic diseases such as (Cd, Hg, Pb, and Tl). The heavy metals concentration can be determined by atomic absorption photometers, spectrophotometer, or inductively coupled plasma (ICP) for very low concentration.
- Nonmetallic compounds: This group includes nitrates (NO_3^-) and cyanides (CN^-). Regarding cyanide, as Mackenzie stated it causes oxygen deprivation by binding the hemoglobin sites and prevents the red blood cell from carrying the oxygen. This causes a blue skin color syndrome, which is called cyanosis. It also causes chronic effects on the central nervous system and thyroid. Cyanide is normally measured by colorimetric, titrimetric, or electrometric methods.

Toxic Organic Substances

There are more than 100 compounds in water that have been listed in the literature as toxic organic compounds. They will not be found naturally in water; they are usually man-made pollutants. These compounds include insecticides, pesticides, solvents, detergents, and disinfectants. They are measured by highly sophisticated instrumental methods, namely, gas chromatographic (GC), high-performance liquid chromatographic (HPLC), and mass spectrophotometric.

Radioactive Substances

Potential sources of radioactive substances in water include wastes from nuclear power plants, industries, or medical research using radioactive chemicals and mining of uranium ores or other radioactive materials. When radioactive substances decay, they release beta, alpha, and gamma radiation. Exposure of humans and other living things to radiation can cause genetic and somatic damage to the living tissues.

Radon gas is of a great health concern because it occurs naturally in groundwater and is a highly volatile gas, which can be inhaled during the showering process. For drinking water, there are established standards commonly used for alpha particles, beta particles, photons emitters, radium-226 and -228, and uranium.

The unit of radioactivity used in water quality applications is the picocurie per liter (pCi/L); 1 pCi is equivalent to about two atoms disintegrating per minute. There are many sophisticated instrumental methods to measure it.

Biological Parameters of Water Quality

One of the most helpful indicators of water quality may be the presence or lack of living organisms. Biologists can survey fish and insect life of natural waters and assess the water quality on the basis of a computed species diversity index (SDI); hence, a water body with a large number of well-balanced species is regarded as a healthy system. Some organisms can be used as an indication for the existence of pollutants based on their known tolerance for a specified pollutant.

Microorganisms exist everywhere in nature. Human bodies maintain a normal population of microbes in the intestinal tract; a big portion of which is made up of coliform bacteria. Although there are millions of microbes per milliliter in wastewater, most of them are harmless. It is only harmful when wastewater contains wastes from people infected with diseases that the presence of harmful microorganisms in wastewater is likely to occur.

Bacteria

Bacteria are considered to be single-celled plants because of their cell structure and the way they ingest food. Bacteria occur in three basic cell shapes: rod-shaped or bacillus, sphere-shaped or coccus, and spiral-shaped or spirellus. In less than 30 min, a single bacterial cell can mature and divide into two new cells.

Under favorable conditions of food supply, temperature, and pH, bacteria can reproduce so rapidly that a bacterial culture may contain 20 million cells per milliliter after just 1 day. This rapid growth of visible colonies of bacteria on a suitable nutrient medium makes it possible to detect and count the number of bacteria in water.

There are several distinctions among the various species of bacteria. One distinction depends on how they metabolize their food. Bacteria that require oxygen for their metabolism are called aerobic bacteria, while those live only in an oxygen-free environment are called anaerobic bacteria. Some species called facultative bacteria can live in either the absence or the presence of oxygen.

At low temperatures, bacteria grow and reproduce slowly. As the temperature increases, the rate of growth and reproduction doubles in every additional 10 °C (up to the optimum temperature for the species). The majority of the species of bacteria having an optimal temperature of about 35 °C.

A lot of dangerous waterborne diseases are caused by bacteria, namely, typhoid and paratyphoid fever, leptospirosis, tularemia, shigellosis, and cholera. Sometimes, the absence of good sanitary practices results in gastroenteritis outbreaks of one or more of those diseases.

Algae

Algae are microscopic plants, which contain photosynthetic pigments, such as chlorophyll. They are autotrophic organisms and support themselves by converting inorganic materials into organic matter by using energy from the sun, during this process they take in carbon dioxide and give off oxygen. They are also important for wastewater treatment in stabilization ponds. Algae are primarily nuisance organisms in the water supply because of the taste and odor problems they create. Certain species of algae cause serious environmental and public health problems; for example, blue-green algae can kill cattle and other domestic animals if the animals drink water containing those species.

Viruses

Viruses are the smallest biological structures known to contain all genetic information necessary for their own reproduction. They can only be seen by a powerful electronic microscope. Viruses are parasites that need a host to live. They can pass through filters that do not permit the passage of bacteria. Waterborne viral pathogens are known to cause infectious hepatitis and poliomyelitis. Most of the waterborne viruses can be deactivated by the disinfection process conducted in the water treatment plant.

Protozoa

Protozoa are single-celled microscopic animal, consume solid organic particles, bacteria, and algae for food, and they are in turn ingested as food by higher level multicellular animals. Aquatic protozoa are floating freely in water and sometimes called zooplankton. They form cysts that are difficult to inactivate by disinfection.

Indicator Organisms

A very important biological indicator of water and pollution is the group of bacteria called coliforms. Pathogenic coliforms always exist in the intestinal system of humans, and millions are excreted with body wastes. Consequently, water that has been recently contaminated with sewage will always contain coliforms.

A particular species of coliforms found in domestic sewage is Escherichia coli or E. coli. Even if the water is only slightly polluted, they are very likely to be found. There are roughly 3 million of E. coli bacteria in 100 mL volume of untreated sewage. Coliform bacteria are aggressive organisms and survive in the water longer than most pathogens. There are normally two methods to test the coliform bacteria—the membrane filter method and multiple-tube fermentation method.

Testing for Coliforms: Membrane Filter Method

A measured volume of sample is filtered through a special membrane filter by applying a partial vacuum. The filter, a flat paper-like disk, has uniform microscopic pores small enough to retain the bacteria on its surface while allowing the water to pass through. The filter paper is then placed in a sterile container called a petri dish, which contains a special culture medium that the bacteria use as a food source.

Then, the petri dish is usually placed in an incubator, which keeps the temperature at 35 °C, for 24 h. After incubation, colonies of coliform bacteria each containing millions of organisms will be visible. The coliform concentration is obtained by counting the number of colonies on the filter; each colony counted represents only one coliform in the original sample. Coliform concentrations are expressed in terms of the number of organisms per 100 mL of water as follows:

$$\text{Coliforms per 100 mL} = \text{Number of colonies} \times 100 / \text{mL of sample}$$

Water Quality Requirements

Water quality requirements differ depending on the proposed used of water. "Water unsuitable for one use may be quite satisfactory for another and water may be considered acceptable for a particular use if water of better quality is not available".

Water quality requirements should be agreed with the water quality standards, which are put down by the governmental agency and represent the legislation requirements. In general, there are three types of standards: in-stream, potable water, and wastewater effluent, each type has its own criteria by using the same methods of measurement. The World Health Organization (WHO) has established minimum standards for drinking water that all countries are recommended to meet.

Properties of Groundwater

The geological nature of the soil determines the chemical composition of the groundwater. Water is constantly in contact with the ground in which it stagnates or circulates, so equilibrium develops between the composition of the soil and that of the water: i.e. water that circulates in a sandy or granitic substratum is acidic and has a few minerals. Water that circulates in limestone contains bicarbonates alkalinity.

Some of the most typical characteristics of groundwater are weak turbidity, a constant temperature and chemical composition and almost overall absence of oxygen. Circulating groundwater can have extreme variation in the composition with the appearance of pollutants and various contaminants. Furthermore, groundwater is often very pure microbiologically.

Movement

Groundwater is in constant motion, although the rate at which it moves is generally slower than it would move in a stream because it must pass through the intricate passageways between free spaces in the rock. First groundwater moves downward due to the pull of gravity. But it can also move upwards because it will flow from higher-pressure areas to lower pressure areas.

The rate of groundwater flow is controlled by two properties of the rock: porosity and permeability. Porosity is the percentage of the volume of the rock that is open space (pore space). This determines the amount of water that a rock can contain. In sediments or sedimentary rocks the porosity depends on grain size, grain shapes, the degree of sorting, and the degree of cementation. Permeability is a measure of the degree to which the pore spaces are interconnected, and the size of the interconnections. Low porosity usually results in low permeability, but high porosity does not necessarily imply high permeability. It is possible to have a highly porous rock with little or no interconnections between pores. A good example of a rock with high porosity and low permeability is a vesicular volcanic rock, where the bubbles that once contained gas give the rock a high porosity, but since these holes are not connected to one another the rock has low permeability.

Discharges and Velocity

The rate at which groundwater moves through the saturated zone depends on the permeability of the rock and the hydraulic gradient. The hydraulic gradient is defined as the difference in elevation divided by the distance between two points on the water table. Velocity, V, is then:

$$V = K(h_2 - h_1)/L$$

Where, K is the coefficient of permeability. If we multiply this expression by the area, A, through which the water is moving, then we get the discharge Q (Darcy's Law).

$$Q = AK(h_2 - h_1)/L.$$

Springs and Wells

A spring is an area on the surface of the Earth where the water table intersects the surface and water flows out of the ground. Springs occur when an impermeable rock (called an aquiclude) intersects a permeable rock that contains groundwater (an aquifer). The occurrence of springs is closely related to the geology of an area. If an impervious layer of rock, such as a clay deposit, underlies a layer of saturated soil or rock, then a line of springs will tend to appear on a slope where the clay layer outcrops. Igneous rocks are also impervious to water, yet they are often extensively fractured, and springs commonly appear where these fractures come to the surface. Fractures in limestone are often enlarged by the dissolving action of groundwater, forming small underground channels and caves.

A well is human-made hole that is dug or drilled deep enough to intersect the water table. If the well is dug beneath the water table, water will fill the open space to the level of the water table, and can be drawn out by a bucket or by pumping. An artesian well is a deep drilled well through which water is forced upward under pressure. The geologic conditions necessary for an artesian well are an inclined aquifer sandwiched between impervious rock layers above and below that trap water in it. Water enters the exposed edge of the aquifer at a high elevation and percolates downward through interconnected pore spaces. The water held in these spaces is under pressure because of the weight of water in the portion of the aquifer above it. If a well is drilled from the land surface through the overlying impervious layer into the aquifer, this pressure will cause the water to rise in the well. In areas where the slope of the aquifer is great enough, pressure will drive the water above ground level in a spectacular, permanent fountain.

Water from an artesian well or spring is usually cold and free of organic contaminants, making it desirable for drinking.

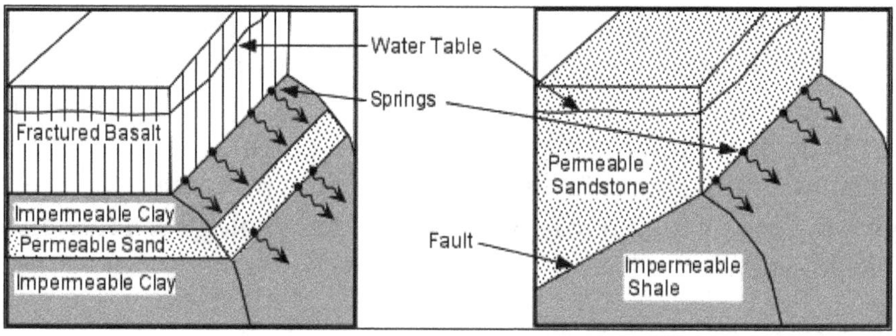

Spring

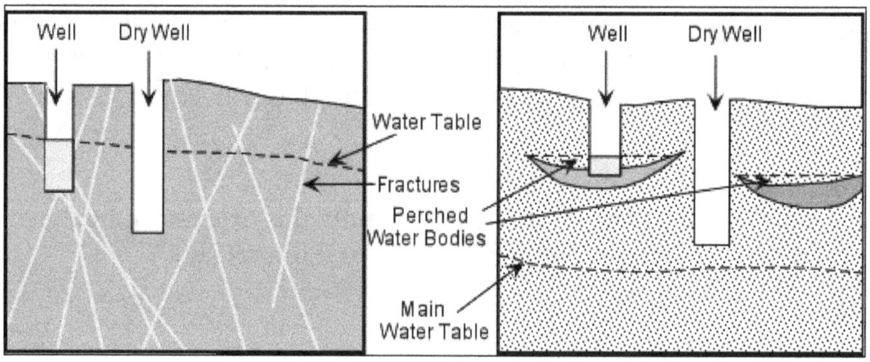

Well

Groundwater Quality

While surface water quality varies due to erosion, runoff, pollution, plant matter, animal waste and other possible contaminants, groundwater quality is generally more stable. And because it usually moves slowly through rock, groundwater is filtered of many contaminants, including some bacteria and viruses. These factors make groundwater an important drinking water resource.

This does not mean, however, that groundwater can't be contaminated: There are many cases worldwide where important groundwater resources (and drinking water supplies) have been ruined by saltwater intrusion (a particular problem in coastal areas), biological contaminants such as manure or septic tank discharge, and industrial chemicals such as pesticides or petroleum products. And once groundwater is contaminated, it's notoriously difficult to remediate.

In addition to concerns about groundwater quality, the overall volume of groundwater in aquifers is also a serious environmental issue. Groundwater is a finite resource, and even large aquifers can be drained of much of their water, especially during droughts, when aquifers aren't recharged by precipitation.

While groundwater quality is much more than groundwater chemistry which mainly includes chemical constituents of groundwater. It also includes biotic constituents of groundwater like bacteria, physical parameters like color, turbidity, odor etc. The issue here is which parameters define groundwater quality and what is the standard for these parameters. There is no universal definition. We must understand that water quality is defined in perspective of its desired use. Thus, the water quality desired for drinking purpose will need to be assessed for a set of parameters in comparison to the agreed standard norm. Similarly, the water quality desired for irrigation purpose will need to be assessed for a different set of parameters in comparison to the agreed standard norm for irrigation water. Holistically we can assess the water quality in terms of physical, chemical and bacteriological parameters for different uses.

Groundwater Contamination and Pollution

Ground water contamination is nearly always the result of human activity. In areas where population density is high and human use of the land is intensive, ground water is especially vulnerable. Virtually any activity whereby chemicals or wastes may be released to the environment, either intentionally or accidentally, has the potential to pollute ground water. When ground water becomes contaminated, it is difficult and expensive to clean up.

To begin to address pollution prevention or remediation, we must understand how surface waters and ground waters interrelate. Ground water and surface water are interconnected and can be fully understood and intelligently managed only when that fact is acknowledged. If there is a water supply well near a source of contamination, that well runs the risk of becoming contaminated. If there is a nearby river or stream, that water body may also become polluted by the ground water

How Does Ground Water Become Contaminated?

Depending on its physical, chemical, and biological properties, a contaminant that has been released into the environment may move within an aquifer in the same manner that ground water moves. (Some contaminants, because of their physical or chemical properties, do not always follow ground water flow.) It is possible to predict, to some degree, the transport within an aquifer of those substances that move along with ground water flow. For example, both water and certain contaminants flow in the direction of the topography from recharge areas to discharge areas. Soils that are porous and permeable tend to transmit water and certain types of contaminants with relative ease to an aquifer below.

Just as ground water generally moves slowly, so do contaminants in ground water. Because of this slow movement, contaminants tend to remain concentrated in the form of a plume that flows along the same path as the ground water. The size and speed of the plume depend on the amount and type of contaminant, its solubility and density, and the velocity of the surrounding ground water.

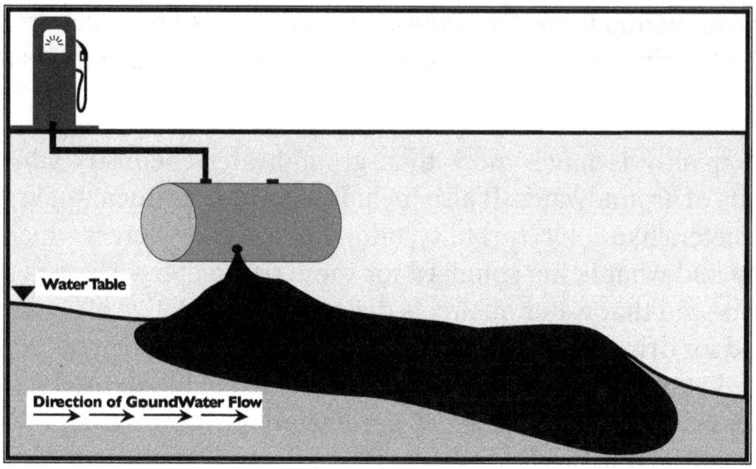

Ground Water Flow.

Ground water and contaminants can move rapidly through fractures in rocks. Fractured rock presents a unique problem in locating and controlling contaminants because the fractures are generally randomly spaced and do not follow the contours of the land surface or the hydraulic gradient. Contaminants can also move into the ground water system through macropores—root systems, animal burrows, abandoned wells, and other systems of holes and cracks that supply pathways for contaminants.

In areas surrounding pumping wells, the potential for contamination increases because water from the zone of contribution, a land area larger than the original recharge area, is drawn into the well and the surrounding aquifer. Some drinking water wells actually draw water from nearby streams, lakes, or rivers. Contaminants present in these surface waters can contribute contamination to the ground water system. Some wells rely on artificial recharge to increase the amount of water infiltrating an aquifer, often using water from storm runoff, irrigation, industrial processes, or treated sewage. In several cases, this practice has resulted in increased concentrations of nitrates, metals, microbes, or synthetic chemicals in the water.

Under certain conditions, pumping can also cause the ground water (and associated contaminants) from another aquifer to enter the one being pumped. This phenomenon is called interaquifer leakage. Thus, properly identifying and protecting the areas affected by well pumping is important to maintain ground water quality.

Generally, the greater the distance between a source of contamination and a ground water source, the more likely that natural processes will reduce the impacts of contamination. Processes such as oxidation, biological degradation (which sometimes renders contaminants less toxic), and adsorption (binding of materials to soil particles) may take place in the soil layers of the unsaturated zone and reduce the concentration of a contaminant before it reaches ground water. Even contaminants that reach ground water directly, without passing through the unsaturated zone, can become less concentrated by dilution (mixing) with the ground water. However, because ground water usually moves slowly, contaminants generally undergo less dilution than when in surface water.

Sources of Ground Water Contamination

Ground water can become contaminated from natural sources or numerous types of human activities. Residential, municipal, commercial, industrial, and agricultural activities can all affect ground water quality. Contaminants may reach ground water from activities on the land surface, such as releases or spills from stored industrial wastes; from sources below the land surface but above the water table, such as septic systems or leaking underground petroleum storage systems; from structures beneath the water table, such as wells; or from contaminated recharge water.

Natural Sources

Some substances found naturally in rocks or soils, such as iron, manganese, arsenic, chlorides, fluorides, sulfates, or radionuclides, can become dissolved in ground water. Other naturally occurring substances, such as decaying organic matter, can move in ground water as particles. Whether any of these substances appears in ground water depends on local conditions. Some substances may pose a health threat if consumed in excessive quantities; others may produce an undesirable odor, taste, or color. Ground water that contains unacceptable concentrations of these substances are not used for drinking water or other domestic water uses unless it is treated to remove these contaminants.

Septic Systems

One of the main causes of ground water contamination in the United States is the effluent (outflow) from septic tanks, cesspools, and privies.

Table: Typical sources of potential ground water contamination by land use category.

Category	Contaminant Source	
Agriculture	Animal burial areas	Irrigation sites
	Animal feedlots	Manure spreading areas/pits
	Fertilizer storage/use	Pesticides storage/use
Commercial	Airports	Jewelry/metal plating
	Auto repair shops	Laundromats
	Boat yards	Medical institutions
	Construction areas	Paint shops
	Car washes	Photography establishments
	Cemeteries	Railroad tracks and yards
	Dry cleaners	Research laboratories
	Gas stations	Scrap and junkyards
	Golf courses	Storage tanks
Industrial	Asphalt plants	Petroleum production/storage
	Chemical manufacture/storage	Pipelines
	Electronics manufacture	Septage lagoons and sludge sites
	Electroplaters	Storage tanks
	Foundries/metal fabricators	Toxic and hazardous spills
	Machine/metalworking shops	Wells (operating/abandoned)
	Mining and mine drainage	Wood preserving facilities
Residential	Fuel	Septic systems, cesspools
	Furniture stripping/refinishing	Sewer lines
	Household hazardous products	Swimming pools (chemical storage)
	Household lawns	

Other	Hazardous waste landfills	Recycling/reduction facilities
	Municipal incinerators	Road deicing operations
	Municipal landfills	Road maintenance depots
	Municipal sewer lines	Storm water drains/basins
	Open burning sites	Transfer stations

Approximately one-fourth of all homes in the United States rely on septic systems to dispose of their human wastes. Although each individual system releases a relatively small amount of waste into the ground, the large number and widespread use of these systems makes them a serious contamination source. Septic systems that are improperly sited, designed, constructed, or maintained can contaminate ground water with bacteria, viruses, nitrates, detergents, oils, and chemicals. Along with these contaminants are the commercially available septic system cleaners containing synthetic organic chemicals (such as 1,1,1- trichloroethane or methylene chloride). These cleaners can contaminate water supply wells and interfere with natural decomposition processes in septic systems.

Most, if not all, state and local regulations require specific separation distances between septic systems and drinking water wells. In addition, computer models have been developed to calculate suitable distances and densities.

Improper Disposal of Hazardous

Waste Hazardous waste should always be disposed of properly, that is to say, by a licensed hazardous waste handler or through municipal hazardous waste collection days. Many chemicals should not be disposed of in household septic systems, including oils (e.g., cooking, motor), lawn and garden chemicals, paints and paint thinners, disinfectants, medicines, photographic chemicals, and swimming pool chemicals. Similarly, many substances used in industrial processes should not be disposed of in drains at the workplace because they could contaminate a drinking water source. Companies should train employees in the proper use and disposal of all chemicals used on site. The many different types and the large quantities of chemicals used at industrial locations make proper disposal of wastes especially important for ground water protection.

Releases and Spills from Stored Chemicals and Petroleum Products

Underground and aboveground storage tanks are commonly used to store petroleum products and other chemical substances. For example, many homes have underground heating oil tanks. Many businesses and municipal highway departments also store gasoline, diesel fuel, fuel oil, or chemicals in on-site tanks. Industries use storage tanks to hold chemicals used in industrial processes or to store hazardous wastes for pickup by a licensed hauler. Approximately 4 million underground storage tanks exist in the United States and, over the years, the contents of many of these tanks have leaked and spilled into the environment.

If an underground storage tank develops a leak, which commonly occurs as the tank ages and corrodes its contents can migrate through the soil and reach the ground water. Tanks that meet federal/state standards for new and upgraded systems are less likely to fail, but they are not foolproof. Abandoned underground tanks pose another problem because their location is often unknown. Aboveground storage tanks can also pose a threat to ground water if a spill or leak occurs and adequate barriers are not in place.

Improper chemical storage, sloppy materials handling, and poor-quality containers can be major threats to ground water. Tanker trucks and train cars pose another chemical storage hazard. Each year, approximately 16,000 chemical spills occur from trucks, trains, and storage tanks, often when materials are being transferred. At the site of an accidental spill, the chemicals are often diluted with water and then washed into the soil, increasing the possibility of ground water contamination.

Landfills

Solid waste is disposed of in thousands of municipal and industrial landfills throughout the country. Chemicals that should be disposed of in hazardous waste landfills sometimes end up in municipal landfills. In addition, the disposal of many household wastes is not regulated.

Once in the landfill, chemicals can leach into the ground water by means of precipitation and surface runoff. New landfills are required to have clay or synthetic liners and leachate (liquid from a landfill containing contaminants) collection systems to protect ground water. Older landfills, however, do not have these safeguards. Older landfills were often sited over aquifers or close to surface waters and in permeable soils with shallow water tables, enhancing the potential for leachate to contaminate ground water. Closed landfills can continue to pose a ground water contamination threat if they are not capped with an impermeable material (such as clay) before closure to prevent the leaching of contaminants by precipitation.

Surface Impoundments

Surface impoundments are relatively shallow ponds or lagoons used by industries and municipalities to store, treat, and dispose of liquid wastes. As many as 180,000 surface impoundments exist in the United States. Like landfills, new surface impoundment facilities are required to have liners, but even these liners sometimes leak.

Sewers and other Pipelines

Sewer pipes carrying wastes sometimes leak fluids into the surrounding soil and ground water. Sewage consists of organic matter, inorganic salts, heavy metals, bacteria, viruses, and nitrogen. Other pipelines carrying industrial chemicals and oil brine have also been known to leak, especially when the materials transported through the pipes are corrosive.

Pesticide and Fertilizer use

Millions of tons of fertilizers and pesticides (e.g., herbicides, insecticides, rodenticides, fungicides, avicides) are used annually in the United States for crop production. In addition to farmers, homeowners, businesses (e.g., golf courses), utilities, and municipalities use these chemicals. A number of these pesticides and fertilizers (some highly toxic) have entered and contaminated ground water following normal, registered use. Some pesticides remain in soil and water for many months to many years. Another potential source of ground water contamination is animal wastes that percolate into the ground from farm feedlots. Feedlots should be properly sited and wastes should be removed at regular intervals.

Between 1985 and 1992, EPA's Office of Pesticides and Toxic Substances and Office of Water conducted a National Pesticide Survey to determine the number of drinking water wells nationwide that contain pesticides and nitrates and the concentration of these substances. The survey also analyzed the factors associated with contamination of drinking water wells by pesticides and nitrates. The survey, which included samples from more than 1,300 public community and rural domestic water supply wells, found that approximately 3.6 percent of the wells contained concentrations of nitrates above the federal maximum contaminant level, and that over half of the wells contained nitrates above the survey's minimum reporting limit for nitrate (0.15 mg/L).

The survey also reported that approximately 0.8 percent of the wells tested contained pesticides at levels higher than federal maximum contaminant levels or health advisory levels. Only 10 percent of the wells classified as rural were actually located on farms. There is a higher incidence of contamination by agricultural chemicals in farm wells used for drinking water.

After further analysis, EPA estimated that for the wells that contain pesticides, a significant percentage probably contain chemical concentrations that exceed the federal health-based limits (e.g., maximum contaminant levels or health advisory levels). Approximately 14.6 percent of the wells tested contained levels of one or more pesticides above the minimum reporting limit set in the survey. The most common pesticides found were atrazine and metabolites (breakdown products) of dimethyl tetrachloroterephthalate (DCPA, commonly known as Dacthal), which is used in many utility easement weed-control programs and for lawn care.

Drainage Wells

Drainage wells are used in wet areas to help drain water and transport it to deeper soils. These wells may contain agricultural chemicals and bacteria.

Injection Wells/Floor Drains

Injection wells are used to collect storm water runoff, collect spilled liquids, dispose of wastewater, and dispose of industrial, commercial, and utility wastes. These wells are regulated by the U.S. EPA's Underground Injection Control Program. In New England, these wells may not be used to inject hazardous wastes from industrial, commercial, and utility operations. The injection wells used in this region are typically shallow and include sumps and dry wells used to handle storm water.

Floor drains were historically used by businesses to handle spills. Today, if a business operates or handles waste fluids that drain to a septic system, dry well, or floor drain, it is required to submit information regarding its operation to the U.S. EPA or its state environmental protection agency. Disposal wells that pose threats to drinking water supplies are prohibited and must be closed, connected to a public sewage system, or connected to a storage tank.

Improperly Constructed Wells

Problems associated with improperly constructed wells can result in ground water contamination when contaminated surface or ground water is introduced into the well.

Improperly Abandoned Wells

These wells can act as a conduit through which contaminants can reach an aquifer if the well casing has been removed, as is often done, or if the casing is corroded. In addition, some people use abandoned wells to dispose of wastes such as used motor oil. These wells may reach into an aquifer that serves drinking supply wells. Abandoned exploratory wells (e.g., for gas, oil, or coal) or test hole wells are usually uncovered and are also a potential conduit for contaminants.

Active Drinking Water Supply Wells

Poorly constructed wells can result in ground water contamination. Construction problems, such as faulty casings, inadequate covers, or lack of concrete pads, allow outside water and any accompanying contaminants to flow into the well. Sources of such contaminants can be surface runoff or wastes from farm animals or septic systems. Contaminated fill packed around a well can also degrade well water quality. Well construction problems are more likely to occur in older wells that were in place prior to the establishment of well construction standards and in domestic and livestock wells.

Poorly Constructed Irrigation Wells

These wells can allow contaminants to enter ground water. Often pesticides and fertilizers are applied in the immediate vicinity of wells on agricultural land.

Mining Activities

Active and abandoned mines can contribute to ground water contamination. Precipitation can leach soluble minerals from the mine wastes (known as spoils or tailings) into the ground water below. These wastes often contain metals, acid, minerals, and sulfides. Abandoned mines are often used as wells and waste pits, sometimes simultaneously. In addition, mines are sometimes pumped to keep them dry; the pumping can cause an upward migration of contaminated ground water, which may be intercepted by a well.

Effects of Ground Water Contamination

Contamination of ground water can result in poor drinking water quality, loss of water supply, degraded surface water systems, high cleanup costs, high costs for alternative water supplies, and/or potential health problems.

The consequences of contaminated ground water or degraded surface water are often serious. For example, estuaries that have been impacted by high nitrogen from ground water sources have lost critical shellfish habitats. In terms of water supply, in some instances, ground water contamination is so severe that the water supply must be abandoned as a source of drinking water. In other cases, the ground water can be cleaned up and used again, if the contamination is not too severe and if the municipality is willing to spend a good deal of money. Follow-up water quality monitoring is often required for many years.

Because ground water generally moves slowly, contamination often remains undetected for long periods of time. This makes cleanup of a contaminated water supply difficult, if not impossible. If a cleanup is undertaken, it can cost thousands to millions of dollars.

Once the contaminant source has been controlled or removed, the contaminated ground water can be treated in one of several ways:

- Containing the contaminant to prevent migration.
- Pumping the water, treating it, and returning it to the aquifer.
- Leaving the ground water in place and treating either the water or the contaminant.
- Allowing the contaminant to attenuate (reduce) naturally (with monitoring), following the implementation of an appropriate source control.

Selection of the appropriate remedial technology is based on site-specific factors and often takes into account cleanup goals based on potential risk that are protective of human health and the environment. The technology selected is one that will achieve those cleanup goals. Different technologies are effective for different types of contaminants, and several technologies are often combined to achieve effective treatment. The effectiveness of treatment depends in part on local hydrogeological conditions, which must be evaluated prior to selecting a treatment option.

Given the difficulty and high costs of cleaning up a contaminated aquifer, some communities choose to abandon existing wells and use other water sources, if available. Using alternative supplies is probably more expensive than obtaining drinking water from the original source. A temporary and expensive solution is to purchase bottled water, but it is not a realistic long-term solution for a community's drinking water supply problem. A community might decide to install new wells in a different area of the aquifer. In this case, appropriate siting and monitoring of the new wells are critical to ensure that contaminants do not move into the new water supplies.

Potential Health Problems

A number of microorganisms and thousands of synthetic chemicals have the potential to contaminate ground water. Drinking water containing bacteria and viruses can result in illnesses such as hepatitis, cholera, or giardiasis. Methemo-globinemia or "blue baby syndrome," an illness affecting infants, can be caused by drinking water that is high in nitrates. Benzene, a component of gasoline, is a known human carcinogen. The serious health effects of lead are well known—learning disabilities in children; nerve, kidney, and liver problems; and pregnancy risks. Concentrations in drinking water of these and other substances are regulated by federal and state laws. Hundreds of other chemicals, however, are not yet regulated, and many of their health effects are unknown or not well understood. Preventing contaminants from reaching the ground water is the best way to reduce the health risks associated with poor drinking water quality.

Regulations to Protect Ground Water

Several federal laws help protect ground water quality. The Safe Drinking Water Act (SDWA) established three drinking water source protection programs: the Wellhead Protection Program, Sole Source Aquifer Program, and the Source Water Assessment Program. It also called for regulation of the use of underground injection wells for waste disposal and provided EPA and the states with the authority to ensure that drinking water supplied by public water systems meets minimum health standards. The Clean Water Act regulates ground water that is shown to have a connection with surface water. It sets standards for allowable pollutant discharges to surface water. The Resource Conservation and Recovery Act (RCRA) regulates treatment, storage, and disposal of hazardous and nonhazardous wastes. The Comprehensive Environmental Response, Compensation, and Liability Act (CERCLA, or Superfund) authorizes the government to clean up contamination or sources of potential contamination from hazardous waste sites or chemical spills, including those that threaten drinking water supplies. CERCLA includes a "community right-toknow" provision. The Federal Insecticide, Fungicide, and Rodenticide Act (FIFRA) regulate pesticide use. The Toxic Substances Control Act (TSCA) regulates manufactured chemicals.

Quality Criteria for Irrigation Water

Plants are sensitive to the water quality used for irrigation; the crop yield depends on the type of water used for irrigation purpose. The main parameters identified for assessing water quality for irrigation purposes are: Salinity; Sodium Adsorption Ratio (SAR) and specific ion toxicity (Sodium, chloride, Boron).

Salinity

Very high salinity in irrigation water adversely affects plant growth. The salinity of water is measured by total dissolved solids, generally expressed as mg/l. The total dissolved solids in water control the electrical conductivity of water. Now a days hand held devices are available for estimation of the electrical conductivity of the water. Todd elucidates and approximates Logan and Richard's relationship between TDS and electrical conductivity. It mentions that for water having electrical conductivity in the range of 100 to 5000 micro Siemens/cm the relationship is:

1 mg/l = 1.56 micro Siemens/cm

Since measurement of electrical conductivity is quite easy, most of the time the water quality consideration with respect to salinity is discussed in terms of electrical conductivity. Different agencies and workers have discussed in detail about tolerance of crops vis-à-vis salinity of irrigation water. However, the most holistic consideration with regards to suitability of irrigation water can be taken when salinity is examined vis-à-vis Sodium Adsorption Ratio (SAR). This can be easily done using C-S diagram.

Sodium Adsorption Ratio (SAR)

The sodium content relative to calcium and magnesium content of the irrigation water controls the rate of infiltration of water to the root zone of crops. It has also been elucidated further by Todd that the soil permeability gets reduced by reaction of sodium with the soil.

$$SAR = \frac{Na}{\sqrt{Ca + Mg / 2}}$$

It is to note here that all the cationic concentrations are in meq/liter.

Specific Ion Toxicity

In this category, we have toxicity primarily by chloride, sodium and boron ions. What happens that circumstantially if the concentration of these ions crosses the critical limit of the crop, the yield reduces drastically? The water needs to be assessed vis –a – vis tolerance of specific crops with regards to concentration of these ions in irrigation water.

Graphical Representation of Groundwater Chemical Quality Data

The chemical quality of groundwater is appreciated better by graphical representation. In the very beginning, we must understand that mostly in graphical representation the major ion concentration (that is Na, K, Mg, Ca, CO_3, HCO_3, SO_4, Cl) is used. Further, while plotting these concentrations are mostly in milli equivalents per liter (meq/liter). The graphical representation of groundwater quality data started by using bar graphs, circular diagram etc. However, limitations of these plotting were that they could not represent substantial number of analysis for one synoptic view. Piper improved upon Hill to produce trilinear diagram for representation of major ion chemistry of groundwater. A similar plot with more flexibility was proposed by S.A Durov and reported by Zaporozec. Besides Richards proposed a diagram for classification of irrigation water. The diagram has Electrical conductivity on x-axis and SAR on y-axis and thus it is also referred to as C-S diagram.

Trilinear Diagram (Hill-Piper Plot)

This is one of the most useful diagrams for graphical representation of groundwater quality data. It has unique advantage that large number of data can be plotted on a single diagram. It also helps in ascertaining hydrochemical facies of groundwater. The hydrochemical facies refers to a cation and anion concentration within a distinct zone. Prior to plotting a data, the concentration of major ions is converted in terms of meq/l. Then the cations are tabulated as percentage of total cations and similarly the anions are tabulated as percentage of total anions. The percentage value of each cation is plotted on the cation triangle on the left and percentage value of each anion is plotted on the anion triangle on the right. The data from cation and anion triangles are projected to the central diamond shaped area as shown by dotted lines. In this way, huge number of data can be plotted and the hydrochemical facies variation can be identified.

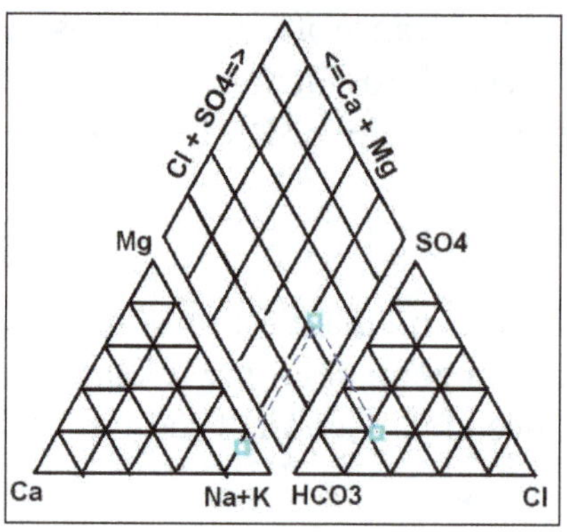

Trilinear diagram for representation of major ion chemistry of groundwater.

Durov's Diagram

The diagram proposed by Durov has again two cation and anion triangles, and in place of central diamond, area there is a square. Like in a Trilinear plot here again major cations and anions are converted in terms of meq/l and then percentage of each action with respect to total cation and percentage of each anion with respect to total anion is tabulated. These percentage values of the cations and anions are then plotted in their respective triangles. In extended Durov's diagram the square can be extrapolated blow and to its right for plotting two more additional parameters.

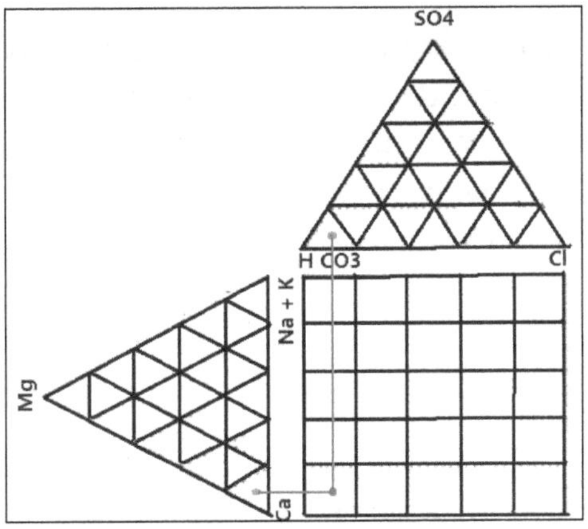

A prototype of Durov's diagram.

C-S Diagram

The C-S diagram is used for classification of irrigation water. Here on logarithmic x-axis, we plot electrical conductivity in terms of micro Siemens/cm and on arithmetic y-axis, we plot SAR. The x-axis helps in assessing salinity hazard and the y-axis helps in assessing sodium hazard. The whole plot area is divided in to low, medium, high and very high category with respect to sodium and salinity hazard. The location of the sample on plot helps in assessing suitability of water for irrigation purpose. In general, as we move towards higher values of sodium or salinity hazard, the unsuitability of irrigation water increases.

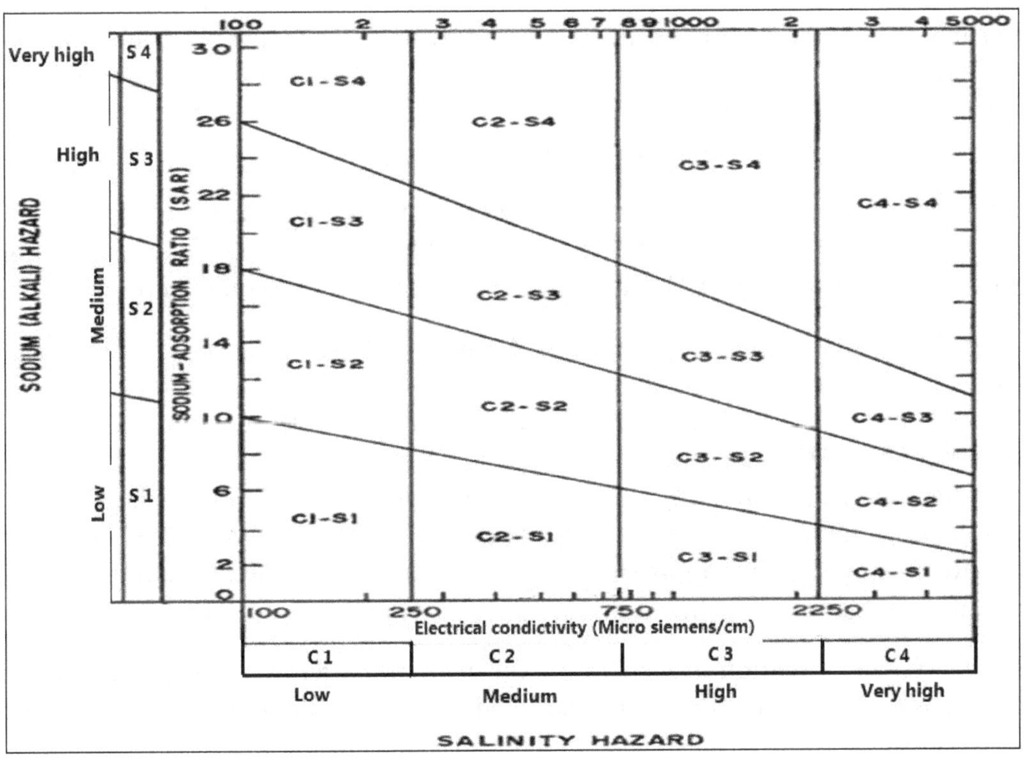

C-S diagram for classification of irrigation water.

Surface Water

Surface water is water on the surface of the planet such as in a river, lake, wetland, or ocean. It can be contrasted with groundwater and atmospheric water. Non-saline surface water uses is replenished by precipitation and by recruitment from ground-water. It is lost through evaporation, seepage into the ground where it becomes ground-water, used by plants for transpiration, extracted by mankind for agriculture, living, industry etc. or discharged to the sea where it becomes saline.

Rivers

A river is a ribbon-like body of water that flows downhill from the force of gravity. A river can be wide and deep, or shallow enough for a person to wade across. A flowing body of water that is smaller than a river is called a stream, creek, or brook. Some rivers flow year-round, while others flow only during certain seasons or when there has been a lot of rain. The largest rivers can be thousands of miles long. The erosional power of rivers can form geologic wonders like the Grand Canyon.

All rivers have a starting point where water begins its flow. This source is called a headwater. The headwater can come from rainfall or snowmelt in mountains, but it can also bubble up from groundwater or form at the edge of a lake or large pond. The other end of a river is called its mouth, where water empties into a larger body of water, such as a lake or ocean. Along the way, rivers may pass through wetlands where plants slow down the water and filter out pollutants.

The water that flows in rivers is fresh, meaning that it contains less than one percent salt. However, rivers still carry and distribute important salts and nutrients to support plant and animal life. For this reason, some of the most biodiverse habitats on our planet can be found around rivers. Collectively, scientists estimate that all the rivers in the world carry about 3.6 billion metric tons (four billion tons) of salt from land to the ocean each year.

Rivers can also form what is called an estuary, where salty seawater mixes with fresh water near the river mouth to form "brackish water." The Hudson River in New York, U.S., is an example of an estuary where brackish water extends more than 241 kilometers (150 miles) upstream.

Fast-flowing rivers carry pebbles, sand, and silt. As the river begins to slow down—as in a wetland, at the outside of a bend, or where the river widens, such as at the mouth–these sediments sink and build up to form deltas. Rivers that overflow their banks also deposit sediment in the surrounding flood plain. These deltas and floodplains are highly fertile agricultural zones that offer tremendous value to the surrounding people. In Egypt, for example, the Nile River and its adjacent delta helped give rise to the Egyptian empire that built the pyramids. Today, farmers in the flood plain of California's Central Valley produce approximately one-third of the vegetables and two-thirds of the fruits and nuts consumed in the United States.

Humans use rivers for irrigation in agriculture, for drinking water, for transportation, to produce electricity through hydroelectric dams, and for leisure activities like swimming and boating. Each of these uses can affect the health of a river and its surrounding ecosystems.

Rivers Serve Many uses

The Chattahoochee River, Atlanta, Georgia, USA. The river serves many purposes, from drinking water to wildlife habitat to a recreation spot for the whole city.

The phrase "river of life" is not just a random set of words. Rivers have been essential not only to humans, but to all life on earth, ever since life began. Plants and animals grow and congregate around rivers simply because water is so essential to all life. It might seem that rivers happen to run through many cities in the world, but it is not that the rivers go through the city, but rather that the city was built and grew up around the river. For humans, rivers are diverted for flood control, irrigation, power generation, public and municipal uses, and even waste disposal.

Where does the Water in Rivers Come from?

Large rivers don't start off large at all, but are the result of much smaller tributaries, creeks, and streams combining, just as tiny capillaries in your body merge to form larger blood-carrying arteries and veins. The mighty river featured in this image is called the Yarlung Tsangpo as it courses through the Tibetan Autonomous Region of China.

The most simplistic answer is that all the water in a river comes from the sky—and that is certainly true, as streamflow is one part of the water cycle. It is also true that most of the water flowing in rivers comes from precipitation runoff from the surrounding landscape (watershed).

But, the water in a river doesn't all come from surface runoff. Rain falling on the land also seeps into the Earth to form groundwater. At a certain depth below the land surface, called the water table, the ground becomes saturated with water. If a river bank happens to cut into this saturated layer, as most rivers do, then water will seep out of the ground into the river. Groundwater seepage can sometimes be seen when water-bearing layers emerge on the land surface, or even on a driveway.

The ground below the water table (the blue area), is saturated, whereas the ground above (the gray area) is not. The top layer (unsaturated soil/rock material) is usually wet, but not totally saturated. Saturated, water-bearing materials often exist in horizontal layers beneath the land surface. Since rivers, in time, may cut vertically into the ground as they flow, the water-bearing layers of rock can become exposed on the river banks. Thus, some of the water in rivers is attributed to flow coming out of the banks. This is why even during droughts there is usually some water in streams.

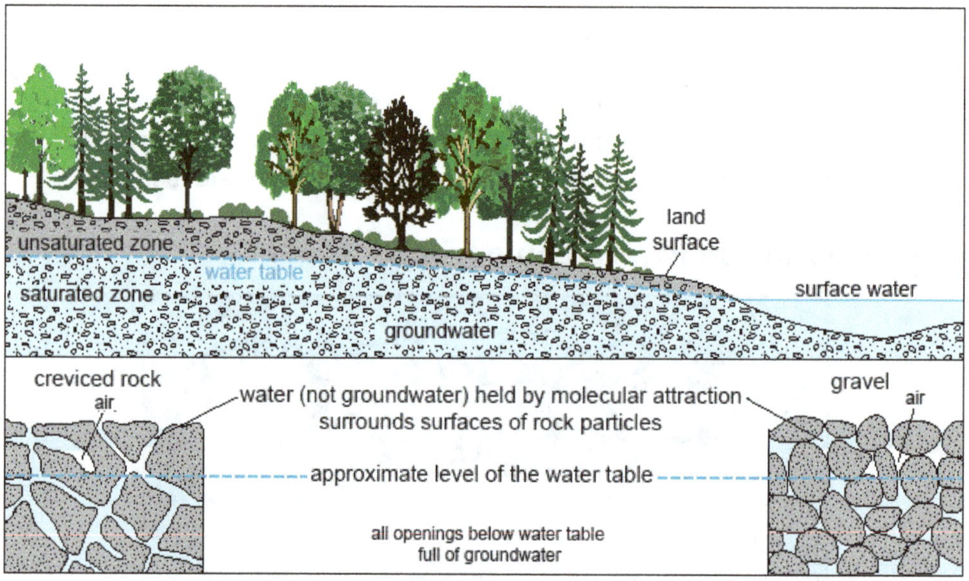

In the diagram you can see how the ground below the water table (the blue area) is saturated with water. The "unsaturated zone" above the water table (the gray area) still contains water (after all, plants' roots live in this area), but it is not totally saturated with water. You can see this in the two drawings at the bottom of the diagram, which show a close-up of how water is stored in-between underground rock particles.

Where does the Water in Rivers Go?

Like everything else on (and in) the Earth, water obeys the rules of gravity and tries to get to the center of the Earth (did you imagine that every molecule in your body is trying to do this, also?). So, the water in rivers flows downhill, with the ultimate goal of flowing into the oceans, which are at sea level. River water may end up in a lake or reservoir, in a pipe aimed at Farmer Joe's corn stalks, in a local swimming pool, or in your drinking glass, but much of it eventually ends up back in the oceans, where it rejoins the water cycle, which is ALWAYS in progress.

Reservoirs

A reservoir is a man-made lake or large freshwater body of water. Many people think of a reservoir as a lake and might even use the words interchangeably. However, the key difference is that reservoirs are artificial and made by humans, while lakes are naturally occurring bodies of water. Reservoirs are great because they provide a supply of water for when naturally occurring bodies of water, like lakes or rivers, run dry.

Types of Reservoirs

There are three main types of reservoirs; valley-dammed reservoirs, bank-side reservoirs, and service reservoirs. Valley-dammed reservoirs are created in valleys between mountains. Usually, there is an existing lake or body of water. The mountain sides are used as the walls of the reservoir to hold the water. A dam, or artificial wall in the reservoir, is built at the narrowest point to hold in the water.

To create a valley-dammed reservoir, the river that will fill the reservoir must be diverted, so the ground can be cleared to lay a foundation for the dam. Next, a concrete lining is put in place, and dam construction can begin. It can take years to build a dam, but once it's done, the water pools in the valley, and a large source of water becomes available.

Bank-side reservoirs are reservoirs that are made by diverting water from local rivers or streams to an existing reservoir. Although this can be applied to many different geographical areas, unlike the valley-dammed reservoir, which requires a valley, diverting water from a river can create problems. The River Thames in London is one example of a bank-side reservoir.

Service reservoirs are entirely man made. They are usually stored in concrete basins above or below ground. You might be familiar with the large water towers in the countryside. This water tower is a service reservoir.

In some areas, people dig cisterns, or service reservoirs that are underground. Cisterns must be in a place of higher elevation to allow the water to flow where it needs to go, whereas water towers are already at a higher location than the surrounding land.

Lakes

Lake is any relatively large body of slowly moving or standing water that occupies an inland basin of appreciable size. Definitions that precisely distinguish lakes, ponds, swamps, and even rivers and other bodies of non-oceanic water are not well established. It may be said, however, that rivers and streams are relatively fast moving; marshes and swamps contain relatively large quantities of grasses, trees, or shrubs; and ponds are relatively small in comparison with lakes. Geologically defined, lakes are temporary bodies of water.

Lake Ann in North Cascades National Park, Washington: Lake Ann in North Cascades National Park, Washington, U.S., viewed from the park's Maple Loop Trail. The North Cascades National Park is a large wilderness area that preserves majestic mountain scenery, snowfields, glaciers, and other unique natural features.

Occurrence

Within the global hydrologic cycle, freshwater lakes play a very small quantitative role, constituting only about 0.009 percent of all free water, which amounts to less than 0.4 percent of all continental fresh water. Saline lakes and inland seas contain another 0.0075 percent of all free water. Freshwater lakes, however, contain well over 98 percent of the important surface waters available for use. Apart from that contained in saline bodies, most other continental waters are tied up in glaciers and ice sheets and the remainder is in groundwater.

Four-fifths of the 125,000 cubic km (30,000 cubic miles) of lake waters occurs in a small number of lakes, perhaps 40 in all. Among the largest are Lake Baikal, in Central Asia, containing about 23,000 cubic km (5,500 cubic miles) of water; Lake Tanganyika (19,000 cubic km [4,600 cubic miles]), in eastern Africa; and Lake Superior (12,000 cubic km [2,900 cubic miles]), one of the Great Lakes of North America. The Great Lakes contain a total of about 25,000 cubic km (6,000 cubic miles) of water and, together with other North American lakes larger than 10 cubic km (2 cubic miles), constitute about one-fourth of the world's lake waters. The Caspian Sea, though not considered a lake by some hydrologists, is the world's largest inland sea. Located in Central Asia, the Caspian Sea has an area of about 386,000 square km (149,000 square miles).

Although lakes are to be found throughout the world, the continents of North America, Africa, and Asia contain about 70 percent of the total lake water, the other continents being less generously endowed. Lakes also occur far beneath the ice sheets of Antarctica; however, surveys of the volume and other features of those discovered so far remain incomplete. One-fourth of the total volume of lake water is spread throughout the world in uncounted numbers of small lakes. Anyone who has flown over much of the Canadian plains area cannot help but be struck by the seemingly endless skein of lakes and ponds covering the landscape below. Though the total volume of water involved is comparatively small, the surface area of lake water is substantial. The total surface area of all Canadian lakes has been estimated to exceed the total surface area of the province of Alberta. The U.S. state of Alaska has more than three million lakes with surface areas greater than 8 hectares (20 acres).

The larger, deeper lakes are a significant factor within the cycle of water—from rain to surface water, ice, soil moisture, or groundwater and thence to water vapour. These lakes receive the drainage from vast tracts of land, store it, pass it on seaward, or lose it to the atmosphere by evaporation. On a local basis, even the smaller lakes play an important hydrologic role. The relatively high ratio of exposed surface area to the total water volume of these lakes accentuates their effectiveness as evaporators. In some cases the efficiency of lakes in losing water to the atmosphere is locally undesirable, because of public and industrial requirements for lake water. A striking example of this condition is the Aral Sea, located in Central Asia. Although it is still one of the world's largest bodies of inland water, in the second half of the 20th century its area was reduced by two-fifths and its mean surface level had dropped by more than 12 metres (40 feet), primarily as a result of the diversion of the Syr Darya and Amu Darya rivers for irrigating adjoining fields. In some basins (e.g., the Chad basin in Africa), lakes are the terrestrial end point of the hydrologic cycle. With no outflow downstream toward the oceans, these closed lakes swell or recede according to the balance of local hydrologic conditions.

Uses and Abuses of Lakes

In today's industrial societies, requirements for water—much of which is derived from lakes—include its use for dilution and removal of municipal and industrial wastes, for cooling purposes, for irrigation, for power generation, and for local recreation and aesthetic displays. Obviously, these requirements vary considerably among regions, climates, and countries.

In another vein, it is convenient to use water to dilute liquid and some solid wastes to concentrations that are not intolerable to the elements of society that must be exposed to the effluent or wish to use it. The degree of dilution that may be acceptable varies from situation to situation and is often in dispute. In some cases, dilution is used purely to facilitate transport of the wastes to purification facilities. The water may then be made available for reuse.

Lake water is also used extensively for cooling purposes. Although this water may not be affected chemically, its change in thermal quality may be detrimental to the environment into which it is disposed, either directly, by affecting fish health or functions, or indirectly, by causing excessive plant production and ultimate deoxygenation due to biological decay. Both fossil-fueled and nuclear power plants are major users of cooling water. Steel mills and various chemical plants also require large quantities.

Economy and Ecology

Concern with thermal pollution of surface waters is concentrated principally on rivers and small lakes. With power requirements in modern societies increasing by about 7 percent per year, however, some apprehension has been expressed about the future thermal loading of even the largest lakes. It was predicted that thermal inputs to each of the North American Great Lakes would increase by nearly 11 times during the last three decades of the 20th century. In terms of energy to be disposed in this fashion, the numbers are staggeringly large. These lakes have such large volumes, however, and such large surface areas (from which much of the heat goes into the atmosphere) that there is some question about the nature and magnitude of the actual effects.

The economic importance of waterways as communication links is enormous. In the earliest times, when travel by many societies was substantially by water, travel routes became established that resulted in relationships between cultural factors and surface hydrology networks. Today river and lake systems serve as communication links and play an important role in shipping because of the large cargo capacities of merchant vessels and the still fairly uncongested condition of inland waterways. Oceanic shipping lanes play the major role, but river and lake systems, which link inland ports with the oceans, have been key factors in the rates of economic growth of many large inland ports.

Commercial fisheries and other food industries reap great harvests from the major lakes of the world. The quality of the fish catch has steadily decreased, however, as a result of pollution in many lakes, with the more desirable species becoming less plentiful and the less desirable species gradually dominating the total. Other commercial harvests from lakes include waterfowl, fur-bearing mammals, and some plant material, such as rice.

Each of the uses described has associated with it the means for abuse of the very characteristics of lakes that make them desirable. Wise management of natural resources has never been humankind's forte. Municipalities and industries have polluted lakes chemically and thermally, the shipping that plies large inland water bodies leaves oil and other refuse in its wake, water used for irrigation often contains chemical residues from fertilizers and biocides when it is returned to lakes, and the populace that so desperately demands clean bodies of water for its recreation often ignores basic sanitary and antipollution practices, to the ultimate detriment of the waters enjoyed.

Problems and Effects

Among the major problems affecting the optimum utilization and conservation of lake waters are eutrophication (aging processes), chemical and biological poisoning, and decreases in water volumes. In the former case, the enrichment of lakes with various nutrients supports biological productivity to an extent in which the ultimate death and decay of biological material places an excessive demand on the oxygen content, resulting in oxygen depletion in the worst cases. Phosphates and nitrates are two of the types of nutrients that are most important in this connection, particularly since they are often introduced in critical quantities in waste effluents from human sources. Other examples of chemical pollution of lakes include the introduction of DDT and other pesticides and heavy metals such as mercury. Bacteriological contamination of lake waters resulting in levels that constitute a hazard to health is another common result of disregard for the environment.

Water-quantity problems are complex, being related to natural vagaries of supply and levels of consumptive utilization of water. In the latter case, the percentage of water returned to the source after utilization varies with the use. The largest losses are due to actual water diversions and processes that result in evaporative losses. The use of large quantities of lake water for cooling purposes by industry and utilities, for example, may raise lake temperatures near the effluents sufficiently to cause increased evaporation. The use of certain types of cooling towers results in even larger losses. Some of the water evaporated will stay within the lake basin, but some will be lost from it.

Another example of this type of loss is connected with the possible application of weather-modification techniques to alleviate the heavy lake-effect snowfalls experienced along the lee shores of large lakes in intermediate latitudes. Redistribution of precipitation always raises the possibility of redistribution of water among various basins.

Lake-effect snowfall is just one example of the influence of lakes on local climate. The ability of large bodies of water to store heat during heating periods and to lose it more gradually than the adjacent landmasses during cooling period's results in a modifying influence on the climate. Because of this propensity, a lake cools air passing over it in summer and warms air passing over it in winter. Consequently, the predominantly downwind side of a lake is more influenced by the ameliorating effects of a lake.

In most instances, moisture is also passed to the atmosphere. In summer, lake cooling serves to stabilize the air mass, but winter heating tends to decrease stability. The moisture-laden, unstable winter flows off lakes produce so-called snowbelts, which affect downwind cities. The snowbelts are usually of limited extent, often within about a kilometre of the lakeshore.

Wetlands

A wetland is an area of land that is either covered by water or saturated with water. The water is often groundwater, seeping up from an aquifer or spring. A wetland's water can also come from a nearby river or lake. Seawater can also create wetlands, especially in coastal areas that experience strong tides.

A wetland is entirely covered by water at least part of the year. The depth and duration of this seasonal flooding varies. Wetlands are transition zones. They are neither totally dry land nor totally underwater; they have characteristics of both.

The saturation of wetland soil determines the vegetation that surrounds it. Plants that live in wetlands are uniquely adapted to their watery (hydric) soil. Wetland plants are called hydrophytes. Seasonally dry wetlands or wetlands with slow-moving water can often support trees and other sturdy vegetation. More frequently flooded wetlands have mosses or grasses as their dominant hydrophytes.

Wetlands exist in many kinds of climates, on every continent except Antarctica. They vary in size from isolated prairie potholes to huge salt marshes. They are found along coasts and inland. Some wetlands are flooded woodlands, full of trees. Others are more like flat, watery grasslands. Still others are choked by thick, spongy mosses.

Wetlands go by many names, such as swamps, peatlands, sloughs, marshes, muskegs, bogs, fens, potholes, and mires. Most scientists consider swamps, marshes, and bogs to be the three major kinds of wetlands.

Swamps

A swamp is a wetland permanently saturated with water and dominated by trees. There are two main types of swamps: freshwater swamps and saltwater swamps. Freshwater swamps are common in inland areas. Saltwater swamps protect coasts from the open ocean.

Freshwater Swamps

Freshwater swamps often form on flat land around lakes or streams, where the water table is high and runoff is slow. Seasonal flooding and rainwater cause the water level in these swamps to fluctuate, or change. Water-tolerant plants, such as cattails, lotus, and cypress, grow in the swamp's wet soil. These plants are key to maintaining the swamp's ecosystem. It is common in tropical areas near the Equator. These equatorial swamps usually experience year-round heat and humidity.

The Eastern and Western Congolian Swamp Forests surround the Congo River, in the nations of the Democratic Republic of Congo and the Republic of the Congo. Tall evergreen trees dominate the swamp forests. Many species of these trees, such as bubinga and ovangkol, are harvested for timber. Bubinga and ovangkol are expensive, luxury woods used to make musical instruments such as violins, as well as furniture. The thick canopy of trees means Congolian swamp forests are more shaded and humid than other wetlands. The muddy floor of these swamps is home to hundreds of insects, reptiles, and amphibians, including dozens of species of frogs.

Congolian swamp forests are also home to a wide variety of large mammals. Most of these mammals are herbivores. Colobus and mangabey monkeys eat mostly tropical fruit. Other mammals, such as forest buffalo, forest elephants, and lowland gorillas, feed on the abundant vegetation of the wetland. In fact, an adult male gorilla can eat up to 32 kilograms (45 pounds) of leaves, fruit, and bark every day.

In more temperate climates, cypress trees often grow out of the still waters of freshwater swamps. Spanish moss may hang from tree branches. Willows and other shrubs may grow beneath the trees. Angular knobs called cypress knees sometimes poke as much as 4 meters (13 feet) above the water. Cypress knees are outgrowths of the trees' root systems. Scientists are not sure what purpose knees serve. They may simply provide support, or they may transport oxygen to the roots.

Tiny water plants called duckweed often form a green cover on the surface of the water. Alligators, frogs, and snakes called water moccasins may swim among the plants. Reptiles and amphibians thrive in freshwater swamps because they are adapted to the fluctuating water levels.

Cypress swamps are common throughout the U.S. The bayous of the state of Louisiana, near slow-moving parts of the Mississippi River, are probably the most famous American swamplands. Shrimp, crawfish, wading birds, and fish such as catfish are native to bayous. Distinct cultures have also developed near bayous and other freshwater swamps. In Louisiana, the food and music of Cajun culture is closely associated with bayou wildlife and imagery.

Saltwater Swamps

Saltwater swamps are usually found along tropical coastlines. Formation of these swamps begins with bare flats of mud or sand that are thinly covered by seawater during high tides. The brackish water of saltwater swamps is not entirely seawater, but not entirely freshwater, either.

Some hydrophytes, such as mangrove trees, can tolerate brackish water. Mangroves are easy to recognize because of their tall, stilt-like roots, which hold the small trunks and branches of the trees above water. Mangrove roots anchor sediment and help soil accumulate around them. They also help build sediment through their growth and decay. Many organisms live among mangrove roots. The root system provides shelter and a place to feed on fallen leaves and other material. Crabs, conchs, and other shellfish are abundant in mangrove swamps.

Saltwater swamps are also home to a huge variety of birds. Mangrove roots and branches provide excellent nesting sites. Saltwater swamps are home to seabirds, such as gulls, as well as freshwater birds, such as herons. The abundance of plants, insects, and small animals provides food for these birds, whose droppings help fertilize the swamp.

The Sundarbans, a saltwater swamp in India and Bangladesh, has the largest mangrove forest in the world. Located on mud flats near the delta of the Ganges River, the area is saturated in freshwater. The Sundarbans also experience strong tides from the Indian Ocean. The biodiversity of the Sundarbans stretches from tiny algae and moss to Bengal tigers. The Bangladeshi portion of the wetland is a UNESCO World Heritage Site. Dozens, perhaps hundreds, of different species of mangrove trees thrive in the Sundarbans. In drier areas of the swamp, palms and grasses grow. Insects such as bees build hives in the trees. In fact, harvesting honey has been a major economic activity in the Sundarbans for centuries.

Bees and other insects are one of the main food sources for tropical birds in the area. Storks, ibises, and herons nest in the high branches of mangrove and palm trees. Smaller birds such as kingfishers and pigeons roost in shrubs. Some birds feed on the hundreds of fish that inhabit the Sundarbans' brackish water: rays, carp, eels, crabs, and shrimp.

Many reptiles and amphibians live in and around the swamp, including frogs, toads, turtles, and snakes. Some of the snakes of the Sundarbans, such as the Indian python, regularly grow up to 3 meters (10 feet) long. Monitor lizards and crocodiles, also native to the Sundarbans, are even larger.

The large reptiles of the Sundarbans regularly prey on mammals such as deer, boar, mongooses, and monkeys. However, the most famous predator of the Sundarbans is the Bengal tiger, an endangered species. Bengal tigers are apex predators—human beings are their only natural predator. In the Sundarbans, Bengal tigers swim in the swampy water and climb trees. The cats, which can grow to 220 kilograms (484 pounds), have been known to attack people in the swamp. Scientists and honey collectors are especially at risk.

Marshes

North and south of the tropics, swamps give way to marshes. These wetlands form a flat, grassy fringe near river mouths, in bays, and along coastlines. Many are alternately flooded and exposed by the movement of tides. Like swamps, marshes are often divided into freshwater and saltwater categories.

Freshwater Marshes

Freshwater marshes, often found hundreds of kilometers from the coast, are dominated by grasses and aquatic plants. These marshes often develop around lakes and streams. Many freshwater marshes lie in the prairie pothole region of North America, the heart of which extends from central Canada through the northern Midwest of the United States.

Prairie potholes are bowl-shaped depressions left by chunks of glacial ice buried in the soil during the most recent ice age. When the ice melted, muddy water filled the potholes. Fertile soil and a temperate climate make these marshes some of the richest in the world. For this reason, many prairie potholes have been drained and the land used for agriculture. Thousands of migratory birds depend on the remaining prairie potholes as they travel from the Arctic to more temperate climates every year.

Farther south, freshwater marshes form much of the Everglades, a huge wetland region in southern Florida. Water from Lake Okeechobee flows slowly through the Everglades on its way to the ocean. Sawgrass, cypress, and mangroves grow along its path. The muddy, slow-moving water is also home to rare types of orchid.

The Everglades are known for their diversity of wildlife. This marsh contains hundreds of species of wading birds, each of which is adapted to feed on insects, fish, clams, shrimp, or even rodents such as mice. Alligators make their nests in the dense sawgrass, and swim in the murky water. Deer and the endangered Florida panther live in the dry areas of the marsh, while manatees and even dolphins swim in the so-called "River of Grass".

Saltwater Marshes

Salt marshes are some of the richest ecosystems for biodiversity. Dominated by grasses, they provide food and shelter for algae, fungi, shellfish, fish, amphibians, and reptiles. Wading birds and other animals feed on the vegetation and abundant insects.

The warm saltwater marshes of northern Australia are influenced by the tides of the Indian and Pacific oceans. They often overlap with the freshwater marshes of rivers, such as the Jardine. A few mangrove trees may dot saltwater marshes, but they are dominated by grasses and a layer of algae called an algal mat. This algal mat is home to many insects and amphibians.

A wide variety of birds are found in Australia's saltwater marshes. Some of these birds nest in the shrubs and prey on insects and fish in the area. Others are migratory, only visiting the marsh when their home ranges become too cold or dry to support life.

Australia's saltwater marshes are also home to the saltwater crocodile. These enormous reptiles often spend the wet season in freshwater swamps and rivers, and migrate to saltwater marshes in the dry season. In addition to fish, Australia's saltwater crocodiles hunt larger animals that may pass through the wetland: kangaroos, birds, and wild boars. Many Australian beaches have strict warnings to swimmers during certain seasons, because saltwater crocodiles are a threat to people as well.

Bogs

Swamps and marshes are generally found in warm climates. Bogs are more common in cold or even Arctic areas in North America, Europe, and Asia. They also exist at high altitudes in warmer regions, such as the Sierra Nevada in the United States. Bogs are often called moors or fens in Europe, and muskegs in Canada.

Like many wetlands, bogs develop in areas where the water table, or the upper surface of underground water, is high. They often begin in glacial depressions called kettle lakes, which are deeper than prairie potholes.

A bog forms as a kettle lake gradually fills with plant debris. Leaves, roots, and stems of large plants accumulate on the bed of the lake. As the lake becomes shallower, mosses and other plants growing along the edges of the lake extend into the water. They form a loose, floating layer of tangled vegetation on the water's surface. Eventually, these plants are followed by water-loving grasses and sedges. Soon, the water is choked with vegetation. The oldest, partially decayed vegetation at the bottom of the bog forms a thick, spongy mat called peat.

Peat is a valuable fuel in many parts of the world. It is often the first step in the creation of coal, a fossil fuel. (The fossils in coal are wetland plants.) Some people living near bogs cut and dry squares of peat. It is burned for heating and cooking, or used to insulate buildings. In Ireland, peat supplies a portion of the country's electrical energy.

Bogs preserve more than the remains of plants, however. The bodies of dozens of prehistoric people have been found in bogs in Europe and Asia. These "bog bodies" have been preserved for thousands of years. Bog bodies are in such excellent condition that anthropologists can examine clothes, tattoos, and hair color, and even investigate a cause of death. Most people found in bogs were killed, though historians and anthropologists debate whether they were murdered or sacrificed as part of a religious ritual.

Some bogs can support a person's weight. They are called quaking bogs because the surface quakes when a person walks on the spongy peat. The island of Ireland, with its cool, wet climate, has hundreds of quaking bogs.

Unlike other wetlands, bogs usually are not agriculturally fertile. The amount of acid in the soil and water is generally higher than that in swamps or marshes. The supply of nutrients, especially nitrogen, is low.

Only certain kinds of plants can grow in bogs. Some of the few plants harvested in the wet, acidic soil of bogs are cranberries and blueberries. Plants are autotrophs, meaning they are able to create their own food from air, water, and sunlight. Many bog plants have adapted to the poor nutrients in the soil and water by expanding their food source. Pitcher plants and sundew, common in bogs, are carnivorous: They trap and consume insects.

Because of the limited species of plants, bogs do not have the biodiversity common in other types of wetlands. Insects, common in all wetlands, include butterflies and dragonflies. These insects feed on the nectar in bog flowers. Ireland has dozens of native butterflies found in bogs. Birds, such as geese and pheasant, also make their homes in the bog, although it is unusual to find larger animals. In North America, moose are one of the few large animals that thrive in bog habitats. Moose, the largest species of deer, consume aquatic plants such as pond lilies.

Vital Ecosystems

Wetlands are some of the most valuable ecosystems on Earth. They act like giant sponges or reservoirs. During heavy rains, wetlands absorb excess water, limiting the effects of flooding. Wetlands also protect coastal areas from storm surges that can wash away fragile beaches and coastal communities. Saltwater swamps and tidal salt marshes help secure coastal soil and sand.

Wetland ecosystems also act as water-treatment facilities. The plants, fungi, and algae of a wetland filter wastes and purify water. Nitrates and other runoff chemicals often wash into wetlands from urban areas and farms. Organisms there absorb the harmful chemicals. Pollutants not absorbed by plants slowly sink to the bottom, where they are buried in sand and other sediment.

Wetlands, especially marshes and swamps, are home to a wide variety of plant and animal life. Some animals, such as shrimp, live in tidal marshes. Many marine fishes, such as striped bass, enter coastal wetlands to spawn. Bass swim from the ocean and into salt marshes to lay their eggs. When the eggs hatch, the young bass find plenty of food and some protection in the grasses or tree roots. Oysters live in huge reefs in salt marshes. All of these wetlands are home to economically valuable fisheries.

The Chesapeake Bay watershed, on the East Coast of the United States, includes more than 60,000 hectares (1.5 million acres) of wetlands. Near the bay, the ecosystem is a tidal salt marsh. Farther from the Atlantic Ocean, freshwater marshes appear close to the Susquehanna River and its tributaries.

Chesapeake Bay wetlands are home to an extraordinary variety of wildlife. Plants include grasses, wild rice, pond lily, cattail, alder, and button bushes. Trees such as red maple, black gum, river birch, black willow, Atlantic white cedar, and bald cypress grow in the bay's forested wetlands.

Chesapeake Bay wetlands are a major nesting area for the bald eagle, a symbol of the United States. The world's largest population of osprey also makes its home there. Bald eagles and ospreys feed on fish in the Chesapeake Bay. Many migratory birds, including swans and geese, spend winters in the Chesapeake wetlands. Other animals native to the Chesapeake Bay include muskrats, beavers, otters, turtles, frogs, and numerous shellfish, as well as the fox squirrel and bog turtle, which are endangered species.

Economic Importance

Wetlands are economically important to people. They are popular places for recreational activities, such as hunting, hiking, canoeing, and bird-watching. According to the U.S. Fish and Wildlife Service, Americans spend more than $100 billion on wetland-related recreational activities every year.

More than 75% of the fish and shellfish that are commercially harvested worldwide are linked with wetlands. For example, the blue crab harvest from the Chesapeake Bay in 2007 was valued at about $51 million. This crab is the official "state crustacean" of the U.S. state of Maryland, and plays an important part of the state's identity. Images of blue crabs are on thousands of souvenirs, and many Maryland restaurants serve crab cakes. The blue crab is also on Maryland's license plate featuring the Chesapeake Bay.

For most of history, wetlands were looked upon as wastelands. They do not easily support development. The soil is wet, spongy, and difficult to build on. Wetlands are also home to pests, from mosquitoes to alligators.

Until recently, draining wetlands was accepted practice. Drained wetlands provided land for agriculture, housing, industry, schools, and hospitals. The capital of the United States, Washington, D.C., is built on a drained wetland along the Potomac and Anacostia rivers. Almost half of U.S. wetlands have been destroyed for development. Drainage and peat harvesting have destroyed wetlands in Ireland and Scandinavia.

Many fish that depend on wetlands have become rare. Some of these species, such as flounder, trout, and bass, are commercially important. Freshwater and ocean fisheries depend on wetlands to provide habitat for the next generation of fish.

In the early 1970s, governments began recognizing the enormous value of wetlands. To protect the vanishing ecosystems, hunting and fishing licenses were restricted. Living shorelines and other restoration projects encouraged the development of coastal wetlands to protect communities from storm surges. Fines and restrictions on agricultural and industrial runoff reduced the toxic chemicals spilling into wetlands.

In some parts of the world, including the United States, it is now against the law to alter or destroy wetlands. Through management plans and stricter laws, people are trying to protect remaining wetlands and to recreate them in areas where they have been destroyed.

Surface Water vs. Groundwater

Surface water includes the freshwater that is channeled into stream systems, lakes, and wetlands on land. Groundwater, on the other hand, is contained in subterranean aquifers within the rock layers below the water table – the underground boundary that divides the saturated and unsaturated levels of the ground. Groundwater derives primarily from rainfall and snowmelt that infiltrates through the soil and into the bedrock, where, driven by gravity, it collects between particles, fractures, and cavities inside rock layers.

The majority of the Earth's groundwater lies within a half-mile of the surface. When it hits an impermeable subterranean layer – such as massive, nonporous rock or clay – the groundwater pools and may flow laterally along it.

While groundwater might remain ensconced in aquifers for thousands of years, it can also naturally appear at the surface in the form of springs, seeps, and groundwater-fed lakes. This is due to its movement within sedimentary strata or along impervious geologic layers.

While surface waters are widely used all over the world for human needs, groundwater aquifers supply most of the drinking water in the United States. When you use a well system, you are tapping the groundwater supply on your property.

Through runoff, air fallout, and other sources, surface water can contain significant amounts of contaminants such as chemical pollutants. And because groundwater is connected to surface water systems through soil percolation and see page, it is also susceptible to contamination. However, the rock and sediment layers below the water table act as natural filters. Therefore, groundwater generally contains fewer contaminants than surface water and requires less treatment.

Minerals in Water – Groundwater and Hardness

While groundwater typically contains fewer contaminants than your average surface water, it does tend to have higher mineral content due to the dissolving action of water. In fact, water is often referred to as the "universal solvent" because it dissolves more substances than any other liquid. Mineral content is the measure by which we classify water as "hard" or "soft." Hard water contains large amounts of certain ions (primarily calcium and magnesium) while soft water has a lower mineral concentration.

We commonly quantify hardness by measuring the concentration of calcium and magnesium in water and then converting it to the equivalent amount of calcium carbonate, expressing the figure in milligrams per liter (mg/L) of water. The divides water into four different categories based on mineral content, including:

- Soft Water – Zero to 60 mg/L.
- Moderately Hard Water – 61 to 120 mg/L.
- Hard Water – 121 to 180 mg/L.
- Very Hard Water – 180 or more mg/L.

Another way to represent hardness that is often used in the water treatment industry is by grains per gallon (gpg) of calcium and magnesium. Using this measure, the Water Quality Association defines the following hardness levels:

- Soft Water – Zero to 1.0 gpg.
- Slightly Hard – 1.1 to 3.5 gpg.
- Moderately Hard – 3.6 to 7.0 gpg.

- Hard – 7.1 to 10.5 gpg.
- Extremely Hard – 10.6 or more gpg.

Hard water is often undesirable to homeowners because it can leave deposits like lime scale that can clog showerheads, faucets, and plumbing fixtures. Additionally, it's more difficult to work up soap lather if your water supply is too hard.

References

- What-is-groundwater: un-igrac.org, Retrieved 11, April 2020
- Water-quality-parameters, water-quality-science-assessments-and-policy: intechopen.com, Retrieved 15, Jan 2020
- What-is-a-reservoir-definition-formation-characteristics: study.com, Retrieved 19, July 2020
- Groundwater-vs-surface-water: grecoandhaines.com, Retrieved 08, March 2020
- Wetland: nationalgeographic.org, Retrieved 23, Feb 2020

5

Surface Runoff

The flow of water over the surface of Earth due to rain, snowmelt or other sources is known as surface runoff. It occurs when the water cannot rapidly infiltrate in the soil. This chapter has been carefully written to provide an easy understanding of the varied facets of surface runoff such as first flush, meltwater runoff, rainfall runoff, stormwater runoff, urban runoff, etc.

If the amount of water falling on the ground is greater than the infiltration rate of the surface, runoff or overland flow will occur. Runoff specifically refers to the water leaving an area of drainage and flowing across the land surface to points of lower elevation. It is not the water flowing beneath the surface of the ground. This type of water flow is called through flow. Runoff involves the following events:

- Rainfall intensity exceeds the soil's infiltration rate.
- A thin water layer forms that begins to move because of the influence of slope and gravity.
- Flowing water accumulates in depressions.
- Depressions overflow and form small rills.
- Rills merge to form larger streams and rivers.
- Streams and rivers then flow into lakes or oceans.

On a global scale, runoff occurs because of the imbalance between evaporation and precipitation over the Earth›s land and ocean surfaces.. In fact, 86% of the Earth's evaporation occurs over the oceans, while only 14% occurs over land. Of the total amount of water evaporated into the atmosphere, precipitation returns only 79% to the oceans, and 21% to the land. Surface runoff sends 7% of the land based precipitation back to the ocean to balance the processes of evaporation and precipitation.

The distribution of runoff per continent shows some interesting patterns. Areas having the most runoff are those with high rates of precipitation and low rates of evaporation.

Table: Continental runoff

Continent	Runoff Per Unit Area (mm per yr.)
Europe	300
Asia	286

Africa	139
North and Central America	265
South America	445
Australia, New Zealand and New Guinea	218
Antarctica and Greenland	164

Streamflow and Stream Discharge

The term streamflow describes the process of water flowing in the organized channels of a stream or river. Stream discharge represents the volume of water passing through a river channel during a certain period of time. Stream discharge can be expressed mathematically with the following equation:

$$Q = W \times D \times V$$

where,

Q equals stream discharge usually measured in cubic meters per second, W equals channel width, D equals channel depth, and V equals velocity of flowing water.

Because of streamflow's potential hazard to humans many streams are gauged by mechanical recorders. These instruments record the stream's discharge on a hydrograph. The graph below illustrates a typical hydrograph and its measurement of discharge over time.

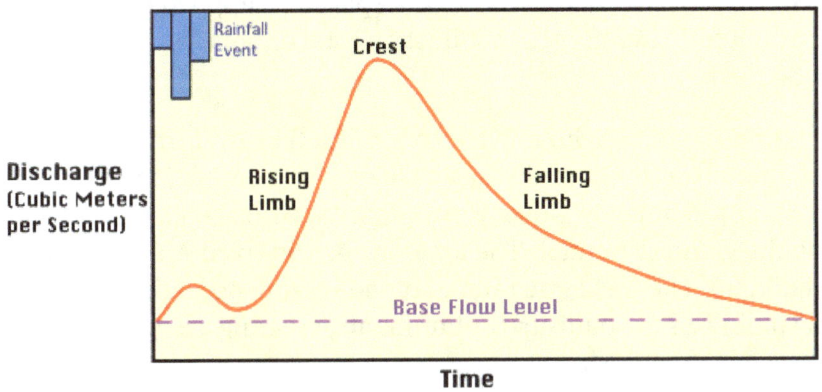

Figure: Stream hydrograph.

From this graph we can observe the following things:

- A small blip caused by rain falling directly into the channel is the first evidence that stream discharge is changing because of the rainfall.
- A significant time interval occurs between the start of rain and the beginning of the main rise in discharge on the hydrograph. This lag occurs because of the time required for the precipitation that falls in the stream's basin to eventually reach the recording station. Usually, the larger the basin the greater the the time lag.
- The rapid movement of surface runoff into the stream›s channels and subsequent flow causes the discharge to rise quickly.

- The falling limb of the hydrograph tends to be less steep that the rise. This flow represents the water added from distant tributaries and from through flow that occurs in surface soils and sediments.
- After some time the hydrograph settles at a constant level known as base flow stage. Most of the base flow comes from groundwater flow which moves water into the stream channel very slowly.

First Flush

The "first flush" phenomenon is generally assumed for rainfall events, and can be described as a concentration first flush or a mass first flush. A concentration first flush occurs when the first runoff has high concentration relative to runoff later in the storm event. A mass first (concentration times flow rate) is flow dependent and it will occurs when both concentration and the initial runoff is high relative to mass emission rate in the later runoff. Concentration first flushes have been frequently reported, but mass first flushes have rarely been quantified. For example, most of the parameters monitored for all the events in this study had higher concentrations at the beginning of the runoff than later in the runoff. Mass first flushes were less frequently observed and with lower magnitudes. This is due to the nature of the runoff, which generally has lower flow rate at the beginning of the storm than in the middle of the storm. Therefore, the mass emission rate in the middle of the storm event may be greater than at the beginning of the storm event, in spite of lower concentrations in the middle of the storm. The concept can be applied to any particular constituent or water quality parameter. Therefore, a first flush in total organic carbon (TOC), for example, can be called a TOC first flush.

The concept of first flush can also be applied to a rainfall season. In California and many other areas of the world, rainfall occurs over distinct periods. For example, the bulk of the rainfall in Los Angeles occurs from approximately November to March, with the months of January and February usually having the greatest rainfall. The long dry period from April or May to October allows contaminants to build up. The first large rainfall of the season, occurring any time from October to January, generally mobilizes the built-up contaminants, creating a larger discharge. This phenomenon is called a "seasonal first flush."

First Flush Effect

The term "first flush effect" refers to rapid changes in water quality (pollutant concentration or load) that occur after early season rains. Soil and vegetation particles wash into streams; sediments and other accumulated organic particles on the river bed are re-suspended, and dissolved substances from soil and shallow groundwater can be flushed into streams. Recent research has shown that this effect has not been observed in relatively pervious areas.

The term is often also used to address the first flood after a dry period, which is supposed to contain higher concentrations than a subsequent one. This is referred to as "first flush flood." There are various definitions of the first flush phenomenon.

First Foul Flush

Storm water runoff in a combined sewer produces a first foul flush with a suspension of accumulated sanitary solids from the sewer in addition to pollutants from surface runoff. Inflow may produce a foul flush effect in sanitary sewers if flows peak during wet weather. As flow rates increase above average, a relatively small percentage of the total flow contains a disproportionately large percentage of the total pollutant mass associated with overall flow volume through the peak flow event. Sewer solids deposition during low flow periods and subsequent resuspension during peak flow events is the major pollutant source for the first-flush combined-sewer overflow (CSO) phenomenon.

Sanitary sewage solids can either go through the system or settle out in laminar flow portions of the sewer to be available for washout during peak flows. The wetted perimeter of sewers may also be colonized by biofilm nourished by soluble sanitary wastes. Hydraulic design is the underlying reason for solids deposition in sewers. Combined sewers sized for peak runoff events expected once a decade can carry up to 1,000 times the average sanitary flow. Less dramatically oversized sewers are common in new developments and near the upstream end of collection systems. Suspended solids may accumulate when low-flow fluid velocities generate insufficient turbulence. Solids deposition is greatest where velocities are low during dry weather. In large combined sewers it may be impossible to attain sanitary sewage velocities generating sufficient turbulence to keep solids suspended during dry weather.

Biofilm and previously deposited solids may be scoured and re-entrained during peak flow turbulence.[1] The high pollution load in wastewater at the beginning of a runoff event occurs when increased flow rate erodes accumulated sewer sediment. Erosion of sediments in sewers can release pollutants in concentrations exceeding levels found in contributing sources. The initial highly polluting foul flush is released at the start of wet weather flow during speedy erosion of a weak layer of highly concentrated surficial sediment bed-load. When conditions favor dry-weather solids deposition, the first foul flush may contain as much as 30 percent of the annual total suspended solids discharged to a combined sewer system. Combined sewer suspended solids concentrations of several thousand milligrams per liter (mg/L) may be observed during the first foul flush.

Pollutant concentration levels are influenced by the age and condition of the collection system and the amount of infiltration/inflow in comparison to the sanitary flow. Pollutant concentration peaks depend on size and slope of the piping system, time interval between storms, and solids accumulation in the collection system. Steeper sewer gradients and pipe bottom shapes that maintain high velocity flow during low-flow conditions will reduce sediment accumulation in sewers; and periodic sewer flushing of individual lines during dry weather may move accumulated solids to the wastewater treatment plant before stormwater runoff causes simultaneous peak flow in the entire collection system.

Meltwater Runoff

As air temperatures rise in the spring, snowpack begins to melt. The water that becomes runoff into streams or recharge into soil is called snowmelt. Most of the snowmelt is a source of water for groundwater recharge. Snowmelt contributes substantial water to soil moisture. Part of the snowmelt that infiltrates into the soil contributes to shallow and deep groundwater. In the late spring the water levels in the river are high because of heavy rainfall experienced at this time. Once rainfall becomes less frequent in the fall, the snowmelt that was stored in the ground will return to the surface and make up the baseflow of the river. In mountains, snowmelt primarily occurs from May to July. Snowmelt occurs earlier in the prairie regions.

Snowmelt

As air temperatures rise in the spring, snowpack begins to melt. The water that becomes runoff into streams or recharge into soil is called snowmelt. Most of the snowmelt is a source of water for groundwater recharge. Snowmelt contributes substantial water to soil moisture. Part of the snowmelt that infiltrates into the soil contributes to shallow and deep groundwater. In the late spring the water levels in the river are high because of heavy rainfall experienced at this time. Once rainfall becomes less frequent in the fall, the snowmelt that was stored in the ground will return to the surface and make up the baseflow of the river. In mountains, snowmelt primarily occurs from May to July. Snowmelt occurs earlier in the prairie regions.

Figure: Snowmelt. When snowpack melts in the spring, some of the snowmelt enters streams, increasing the amount of water in the stream.

Snowmelt and Flooding

The effect of snowmelt on potential flooding, mainly during the spring, is something that causes concern for many people around the world. Besides flooding, rapid snowmelt can trigger landslides and debris flows. In alpine regions like Switzerland, snowmelt is a major component of runoff. In combination with specific weather conditions, such as excessive rainfall on melting snow for example, it may even be a major cause of floods. In Switzerland, snowmelt forecasting is being used as a flood-warning tool to predict snowmelt runoff and potential flooding.

In some parts of the world, such as in the Pacific Northwest of the United States, annual springtime flood events occur when rain falls on existing snowpacks, known as a "rain-on-snow event." Runoff during rain-on-snow events has been associated with mass-wasting of hill slopes, damage to riparian (areas alongside streams) zones, downstream flooding and associated damage, and loss of life. Some studies suggest that the amount of forest cover can have an influence on the magnitude of rain-on-snow events.

Hetch-Hetchy basin near Yosemite, California. Photo by David Gay

Figure: Snowmelt in the Hetch-Hetchy basin near Yosemite, California.

Conodoguinet Creek, Pennsylvania. NOAA, Jason Nolan

Figure: Snowmelt in the Conodoguinet creek, Pennsylvania.

Rainfall Runoff

Runoff is generated by rainstorms and its occurrence and quantity are dependent on the characteristics of the rainfall event, i.e. intensity, duration and distribution. There are, in addition, other important factors which influence the runoff generating process.

Rainfall Characteristics

Precipitation in arid and semi-arid zones results largely from convective cloud mechanisms producing storms typically of short duration, relatively high intensity and limited areal extent. However, low intensity frontal-type rains are also experienced, usually in the winter season. When most precipitation occurs during winter, as in Jordan and in the Negev, relatively low-intensity rainfall may represent the greater part of annual rainfall.

Rainfall intensity is defined as the ratio of the total amount of rain (rainfall depth) falling during a given period to the duration of the period It is expressed in depth units per unit time, usually as mm per hour (mm/h).

The statistical characteristics of high-intensity, short-duration, convective rainfall are essentially independent of locations within a region and are similar in many parts of the world. Analysis of short-term rainfall data suggests that there is a reasonably stable relationship governing the intensity characteristics of this type of rainfall. Studies carried out in Saudi Arabia (Raikes and Partners 1971) suggest that, on average, around 50 percent of all rain occurs at intensities in excess of 20 mm/hour and 20-30 percent occurs at intensities in excess of 40 mm/hour. This relationship appears to be independent of the long-term average rainfall at a particular location.

Variability of Annual Rainfall

Water harvesting planning and management in arid and semi-arid zones present difficulties which are due less to the limited amount of rainfall than to the inherent degree of variability associated with it.

In temperate climates, the standard deviation of annual rainfall is about 10-20 percent and in 13 years out of 20, annual amounts are between 75 and 125 percent of the mean. In arid and semi-arid climates the ratio of maximum to minimum annual amounts is much greater and the annual rainfall distribution becomes increasingly skewed with increasing aridity. With mean annual rainfalls of 200-300 mm the rainfall in 19 years out of 20 typically ranges from 40 to 200 percent of the mean and for 100 mm/year, 30 to 350 percent of the mean. At more arid locations it is not uncommon to experience several consecutive years with no rainfall.

For a water harvesting planner, the most difficult task is therefore to select the appropriate "design" rainfall according to which the ratio of catchment to cultivated area will be determined.

Design rainfall is defined as the total amount of rain during the cropping season at which or above which the catchment area will provide sufficient runoff to satisfy the crop water requirements. If the actual rainfall in the cropping season is below the design rainfall, there will be moisture stress in the plants; if the actual rainfall exceeds the design rainfall, there will be surplus runoff which may result in a damage to the structures.

The design rainfall is usually assigned to a certain probability of occurrence or exceedance. If, for example, the design rainfall with a 67 percent probability of exceedance is selected, this means that on average this value will be reached or exceeded in two years out of three and therefore the crop water requirements would also be met in two years out of three.

The design rainfall is determined by means of a statistical probability analysis.

Probability Analysis

The first step is to obtain annual rainfall totals for the cropping season from the area of concern. In locations where rainfall records do not exist, figures from stations nearby may be used with caution. It is important to obtain long-term records the variability of rainfall in arid and semi-arid areas is considerable. An analysis of only 5 or 6 years of observations is inadequate as these 5 or 6 values may belong to a particularly dry or wet period and hence may not be representative for the long term rainfall pattern.

In the following example, 32 annual rainfall totals from Mogadishu (Somalia) were used for an analysis.

Table: annual rainfall, mogadishu (somalia)

Year	R mm	Year	R mm	Year	R mm	Year	R mm	Year	R mm
1957	484	1964	489	1971	271	1977	660	1983	273
1958	529	1965	498	1972	655	1978	216	1984	270
1959	302	1966	395	1973	371	1979	594	1985	423
1960	403	1967	890	1974	255	1980	544	1986	251
1961	960	1968	680	1975	411	1981	563	1987	533
1962	453	1969	317	1976	339	1982	526	1988	531
1963	633	1970	300						

The next step is to rank the annual totals from table above with m == 1 for the largest and m = 32 for the lowest value and to rearrange the data accordingly.

The probability of occurrence P (%) for each of the ranked observations can be calculated from the equation,

$$P(\%) = \frac{m - 0.375}{N + 0.25} \times 100$$

where:

P = probability in % of the observation of the rank m
m = the rank of the observation
N = total number of observations used

Table: ranked annual rainfall data, mogadishu (somalia)

Year	R	m	P	Year	R	m	P	Year	R	m	P	Year	R	m	P
	mm		%		mm		%		mm		%		mm		%
1961	960	1	1.9	1988	531	11	32.9	1966	395	21	64.0	1986	251	31	95.0
1967	890	2	5.0	1958	529	12	36.0	1973	371	22	67.1	1978	216	32	98.1
1968	680	3	8.1	1982	526	13	39.1	1976	339	23	70.2				
1977	660	4	11.2	1965	498	14	42.2	1969	317	24	73.3				
1972	655	5	14.3	1964	489	15	45.3	1959	302	25	76.4				
1963	633	6	17.4	1957	484	16	48.4	1970	300	26	79.5				
1979	594	7	20.5	1962	453	17	51.6	1983	273	27	82.6				
1981	563	8	23.6	1985	423	18	54.7	1971	271	28	85.7				
1980	544	9	26.7	1975	411	19	57.8	1984	270	29	88.8				
1987	533	10	29.8	1960	403	20	60.9	1974	255	30	91.1				

The above equation is recommended for N = 10 to 100 There are several other, but similar, equations known to compute experimental probabilities.

The next step is to plot the ranked observations (columns 2, 6,10, 14, Table above) against the corresponding probabilities (columns 4, 8,12,16, Table above). For this purpose normal probability paper must be used.

Finally a curve is fitted to the plotted observations in such a way that the distance of observations above or below the curve should be as close as possible to the curve. The curve may be a straight line.

From this curve it is now possible to obtain the probability of occurrence or exceedance of a rainfall value of a specific magnitude. Inversely, it is also possible to obtain the magnitude of the rain corresponding to a given probability.

In the above example, the annual rainfall with a probability level of 67 percent of exceedance is 371 mm, i.e. on average in 67 percent of time (2 years out of 3) annual rain of 371 mm would be equalled or exceeded.

For a probability of exceedance of 33 percent, the corresponding value of the yearly rainfall is 531 mm.

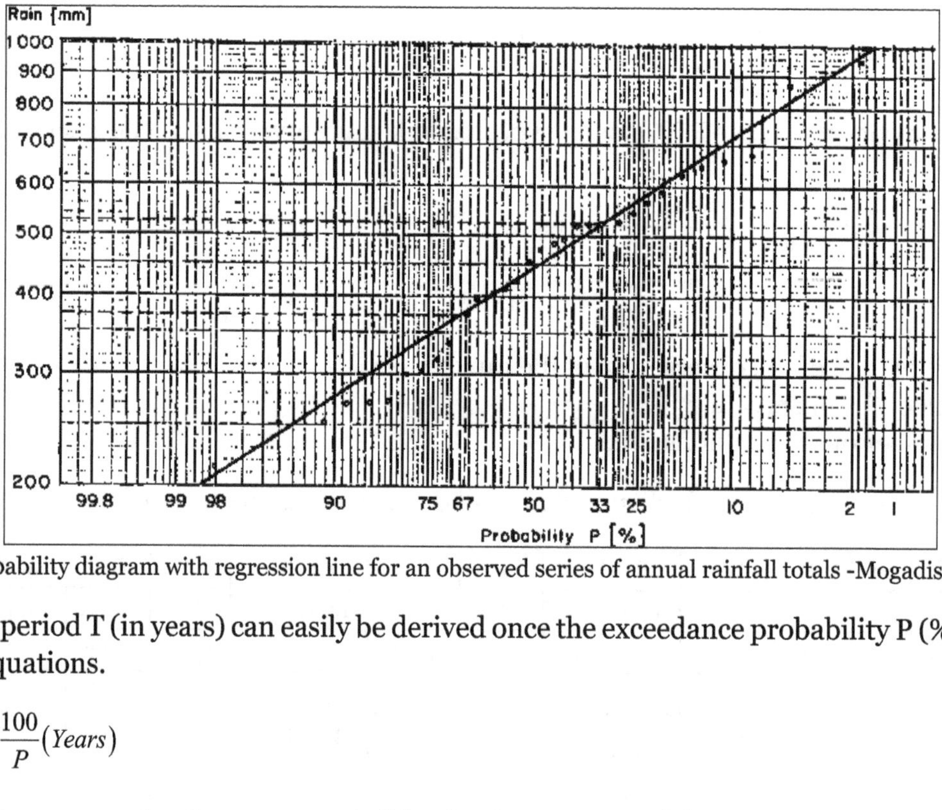

Figure: Probability diagram with regression line for an observed series of annual rainfall totals -Mogadishu, Somalia.

The return period T (in years) can easily be derived once the exceedance probability P (%) is known from the equations.

$$T = \frac{100}{P}(Years)$$

From the above examples the return period for the 67 percent and the 33 percent exceedance probability events would thus be:

$$T_{67\%} = \frac{100}{67} = 1.5(Years)$$

i.e. on average an annual rainfall of 371 mm or higher can be expected in 2 years out of 3;

$$T_{33\%} = \frac{100}{33} = 3(Years)$$

respectively i.e. on average an annual rainfall of 531 mm or more can only be expected in 1 year out of 3.

Rainfall-runoff Relationship

The Surface Runoff Process

When rain falls, the first drops of water are intercepted by the leaves and stems of the vegetation. This is usually referred to as interception storage.

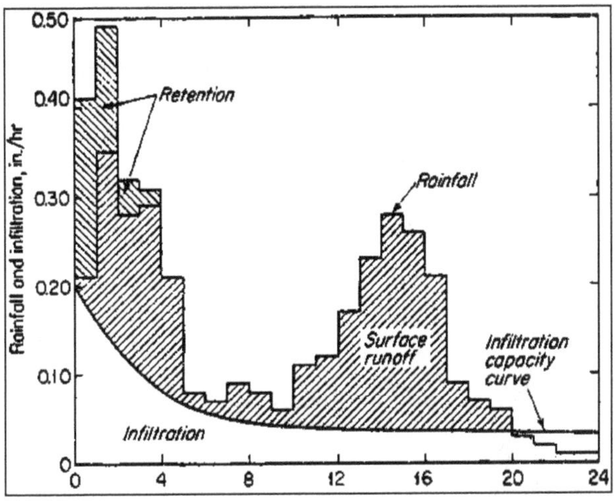

Figure: Schematic diagram illustrating relationship between rainfall, infiltration and runoff.

As the rain continues, water reaching the ground surface infiltrates into the soil until it reaches a stage where the rate of rainfall (intensity) exceeds the infiltration capacity of the soil. Thereafter, surface puddles, ditches, and other depressions are filled (depression storage), after which runoff is generated.

The infiltration capacity of the soil depends on its texture and structure, as well as on the antecedent soil moisture content (previous rainfall or dry season). The initial capacity (of a dry soil) is high but, as the storm continues, it decreases until it reaches a steady value termed as final infiltration rate.

The process of runoff generation continues as long as the rainfall intensity exceeds the actual infiltration capacity of the soil but it stops as soon as the rate of rainfall drops below the actual rate of infiltration.

The rainfall runoff process is well described in the literature. Numerous papers on the subject have been published and many computer simulation models have been developed. All these models, however, require detailed knowledge of a number of factors and initial boundary conditions in a catchment area which in most cases are not readily available.

For a better understanding of the difficulties of accurately predicting the amount of runoff resulting from a rainfall event, the major factors which influence the rainfall-runoff process are described below.

Factors Affecting Runoff

Apart from rainfall characteristics such as intensity, duration and distribution, there are a number of site (or catchment) specific factors which have a direct bearing on the occurrence and volume of runoff.

Soil Type

The infiltration capacity is among others dependent on the porosity of a soil which determines the water storage capacity and affects the resistance of water to flow into deeper layers.

Porosity differs from one soil type to the other. The highest infiltration capacities are observed in loose, sandy soils while heavy clay or loamy soils have considerable smaller infiltration capacities.

Figure below illustrates the difference in infiltration capacities measured in different soil types.

The infiltration capacity depends furthermore on the moisture content prevailing in a soil at the onset of a rainstorm.

The initial high capacity decreases with time (provided the rain does not stop) until it reaches a constant value as the soil profile becomes saturated.

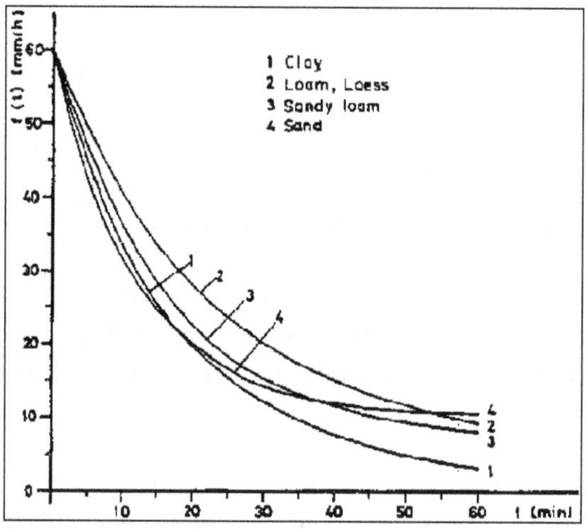

Figure: Infiltration capacity curves for different soil types.

This however, is only valid when the soil surface remains undisturbed.

It is well known that the average size of raindrops increases with the intensity of a rainstorm. In a high intensity storm the kinetic energy of raindrops is considerable when hitting the soil surface. This causes a breakdown of the soil aggregate as well as soil dispersion with the consequence of driving fine soil particles into the upper soil pores. This results in clogging of the pores, formation of a thin but dense and compacted layer at the surface which highly reduces the infiltration capacity.

This effect, often referred to as capping, crusting or sealing, explains why in arid and semi-arid areas where rainstorms with high intensities are frequent, considerable quantities of surface runoff are observed even when the rainfall duration is short and the rainfall depth is comparatively small.

Soils with a high clay or loam content (e.g. Loess soils with about 20% clay) are the most sensitive for forming a cap with subsequently lower infiltration capacities. On coarse, sandy soils the capping effect is comparatively small.

Vegetation

The amount of rain lost to interception storage on the foliage depends on the kind of vegetation and its growth stage. Values of interception are between 1 and 4 mm. A cereal crop, for example, has a smaller storage capacity than a dense grass cover.

The amount of rain lost to interception storage on the foliage depends on the kind of vegetation and its growth stage. Values of interception are between 1 and 4 mm. A cereal crop, for example, has a smaller storage capacity than a dense grass cover.

More significant is the effect the vegetation has on the infiltration capacity of the soil. A dense vegetation cover shields the soil from the raindrop impact and reduces the crusting effect as described earlier.

In addition, the root system as well as organic matter in the soil increase the soil porosity thus allowing more water to infiltrate. Vegetation also retards the surface flow particularly on gentle slopes, giving the water more time to infiltrate and to evaporate.

In conclusion, an area densely covered with vegetation, yields less runoff than bare ground.

Slope and Catchment Size

Investigations on experimental runoff plots have shown that steep slope plots yield more runoff than those with gentle slopes.

In addition, it was observed that the quantity of runoff decreased with increasing slope length.

This is mainly due to lower flow velocities and subsequently a longer time of concentration (defined as the time needed for a drop of water to reach the outlet of a catchment from the most remote location in the catchment). This means that the water is exposed for a longer duration to infiltration and evaporation before it reaches the measuring point. The same applies when catchment areas of different sizes are compared.

The runoff efficiency (volume of runoff per unit of area) increases with the decreasing size of the catchment i.e. the larger the size of the catchment the larger the time of concentration and the smaller the runoff efficiency.

Figure below clearly illustrates this relationship.

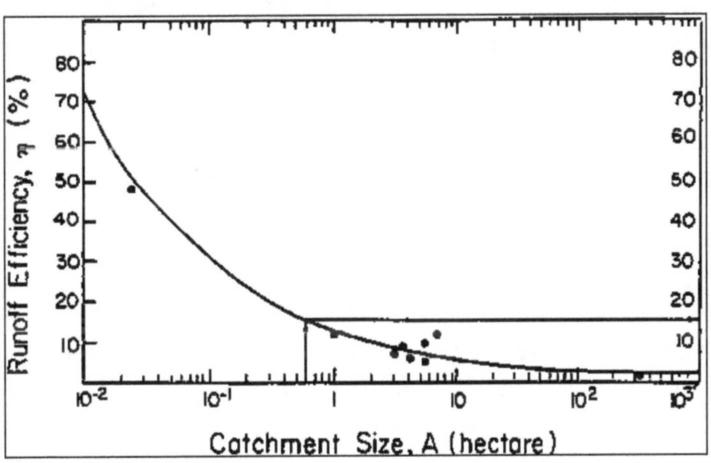

Figure: Runoff efficiency as a function of catchment size (Ben Asher 1988).

It should however be noted that the diagram has been derived from investigations in the Negev desert and not be considered as generally applicable to others regions. The purpose of this diagram is to demonstrate the general trend between runoff and catchment size.

Runoff Coefficients

Apart from the above-mentioned site-specific factors which strongly influence the rainfall-runoff process, it should also be considered that the physical conditions of a catchment area are not homogenous. Even at the micro level there are a variety of different slopes, soil types, vegetation covers etc. Each catchment has therefore its own runoff response and will respond differently to different rainstorm events.

The design of water harvesting schemes requires the knowledge of the quantity of runoff to be produced by rainstorms in a given catchment area. It is commonly assumed that the quantity (volume) of runoff is a proportion (percentage) of the rainfall depth.

Runoff [mm] = K x Rainfall depth [mm]

In rural catchments where no or only small parts of the area are impervious, the coefficient K, which describes the percentage of runoff resulting from a rainstorm, is however not a constant factor. Instead its value is highly variable and depends on the above described catchment-specific factors and on the rainstorm characteristics.

For example, in a specific catchment area with the same initial boundary condition (e.g. antecedent soil moisture), a rainstorm of 40 minutes duration with an average intensity of 30 mm/h would produce a smaller percentage of runoff than a rainstorm of only 20 minutes duration but with an average intensity of 60 mm/h although the total rainfall depth of both events were equal.

Determination of Runoff Coefficients

For reasons explained before, the use of runoff coefficients which have been derived for watersheds in other geographical locations should be avoided for the design of a water harvesting scheme. Also runoff coefficients for large watersheds should not be applied to small catchment areas.

An analysis of the rainfall-runoff relationship and subsequently an assessment of relevant runoff coefficients should best be based on actual, simultaneous measurements of both rainfall and runoff in the project area.

As explained above, the runoff coefficient from an individual rainstorm is defined as runoff divided by the corresponding rainfall both expressed as depth over catchment area (mm):

$$K = \frac{Runoff[mm]}{RainFall[mm]}$$

Actual measurements should be carried out until a representative range is obtained. Shanan and Tadmor recommend that at least 2 years should be spent to measure rainfall and runoff data before any larger construction programme starts. Such a time span would in any case be justified bearing in mind the negative demonstration effect a water harvesting project would have if the structures were seriously damaged or destroyed already during the first rainstorm because the design was based on erroneous runoff coefficients.

When plotting the runoff coefficients against the relevant rainfall depths a satisfactory correlation is usually observed.

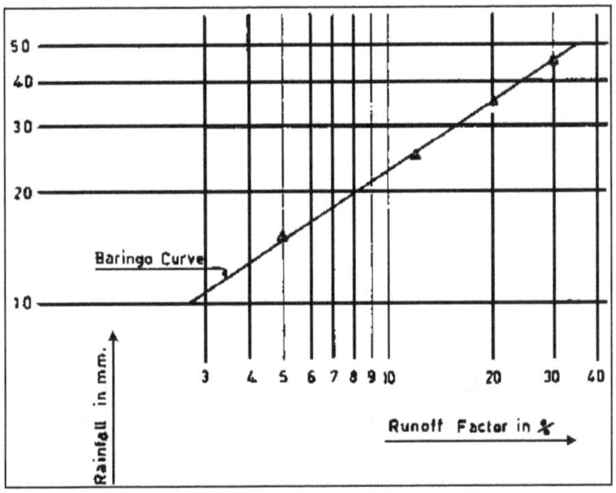

Figure: Rainfall-runoff relationships, Baringo, Kenya.

A much better relationship would be obtained if in addition to rainfall depth the corresponding rainstorm intensity, the rainstorm duration and the antecedent soil moisture were also measured. This would allow rainstorm events to be grouped according to their average intensity and their antecedent soil moisture and to plot the runoff coefficients against the relevant rainfall durations separately for different intensities.

Rainfall intensities can be accurately measured by means of a continuously recording autographic rain gauge.

It is also possible to time the length of individual rainstorms and to calculate the average intensities by dividing the measured rainfall depths by the corresponding duration of the storms.

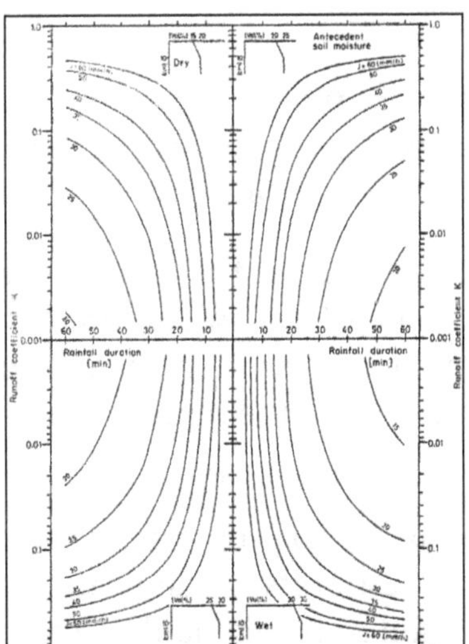

Figure: Runoff coefficients in relation to rainfall intensity, rainfall duration and antecedent soil moisture. Measured on loess soil with sparse vegetation. Ground slope 1.5%.

When analysing the measured data it will be noted that a certain amount of rainfall is always required before any runoff occurs. This amount, usually referred to as threshold rainfall, represents the initial losses due to interception and depression storage as well as to meet the initially high infiltration losses.

The threshold rainfall depends on the physical characteristics of the area and varies from catchment to catchment. In areas with only sparse vegetation and where the land is very regularly shaped, the threshold rainfall may be only in the range of 3 mm while in other catchments this value can easily exceed 12 mm, particularly where the prevailing soils have a high infiltration capacity. The fact that the threshold rainfall has first to be surpassed explains why not every rainstorm produces runoff. This is important to know when assessing the annual runoff-coefficient of a catchment area.

Assessment of Annual or Seasonal Runoff

The knowledge of runoff from individual storms as described before is essential to assess the runoff behaviour of a catchment area and to obtain an indication both of runoff-peaks which the structure of a water harvesting scheme must withstand and of the needed capacity for temporary surface storage of runoff, for example the size of an infiltration pit in a micro catchment system.

However, to determine the ratio of catchment to cultivated area, it is necessary to assess either the annual (for perennial crops) or the seasonal runoff coefficient. This is defined as the total runoff observed in a year (or season) divided by the total rainfall in the same year (or season).

$$k = \frac{Yearly(seasonal)TotalRunoff\left[mm\right]}{Yearly(seasonal)TotalRainFall\left[mm\right]}$$

The annual (seasonal) runoff coefficient differs from the runoff coefficients derived from individual storms as it takes into account also those rainfall events which did not produce any runoff. The annual (seasonal) runoff-coefficient is therefore always smaller than the arithmetic mean of runoff coefficients derived from individual runoff-producing storms.

Runoff Plots

Runoff plots are used to measure surface runoff under controlled conditions. The plots should be established directly in the project area. Their physical characteristics, such as soil type, slope and vegetation must be representative of the sites where water harvesting schemes are planned.

The size of a plot should ideally be as large as the estimated size of the catchment planned for the water harvesting project. This is not always possible mainly due to the problem of storing the accumulated runoff. A minimum size of 3-4 m in width and 10-12 m in length is recommended. Smaller dimensions should be avoided, since the results obtained from very small plots are rather misleading.

Care must be taken to avoid sites with special problems such as rills, cracks or gullies crossing the plot. These would drastically affect the results which would not be representative for the whole area. The gradient along the plot should be regular and free of local depressions. During construction of the plot, care must be taken not to disturb or change the natural conditions of the plot such as destroying the vegetation or compacting the soil. It is advisable to construct several plots in series in the project area which would permit comparison of the measured runoff volumes and to judge on the representative character of the selected plot sites.

Around the plots metal sheets or wooden planks must be driven into the soil with at least 15 cm of height above ground to stop water flowing from outside into the plot and vice versa. A rain gauge must be installed near to the plot. At the lower end of the plot a gutter is required to collect the run-off. The gutter should have a gradient of 1% towards the collection tank. The soil around the gutter should be backfilled and compacted. The joint between the gutter and the lower side of the plot may be cemented to form an apron in order to allow a smooth flow of water from the plot into the gutter. The collection tank may be constructed from stone masonry, brick or concrete blocks, but a buried barrel will also meet the requirements. The tank should be covered and thus be protected against evaporation and rainfall. The storage capacity of the tank depends on the size of the plot but should be large enough to collect water also from extreme rain storms. Following every storm (or every day at a specific time), the volume of water collected in the rain gauge and in the runoff tank must be measured. Thereafter the gauge and tank must be completely emptied. Any silt which may have deposited in the tank and in the gutter must be cleared.

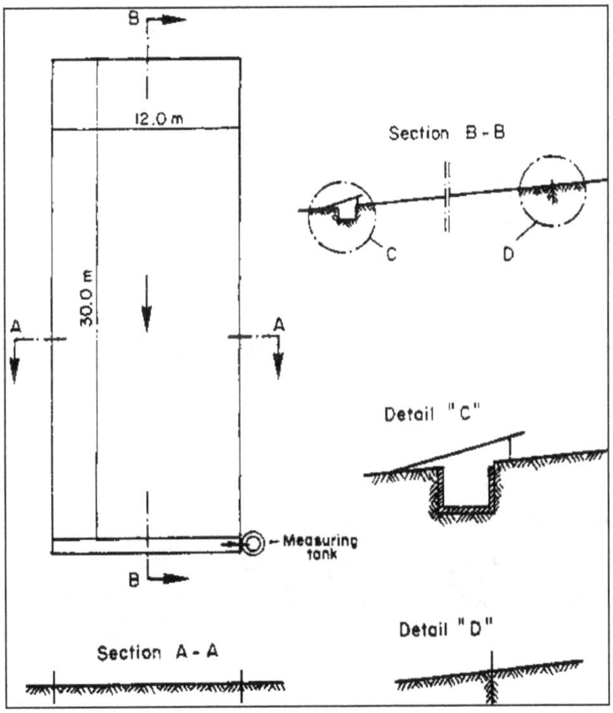

Figure: Standard layout of a runoff plot.

Rainfall-Runoff Modeling

Rainfall-Runoff modeling is one of the most classical applications of hydrology. It has the purpose of simulating the peak river flow or the hydrograph induced by an observed or a hypothetical rain-fall forcing. Rainfall-runoff models may include other input variables, like temperature, information on the catchment or others. Within the context of this subject, we are studying rainfall-runoff models with the purpose of producing estimates of peak river flow, simulation of flood hydrographs or simulation of synthetic river flows in general, even for extended periods, for example for setting up water resources management strategies.

Therefore, rainfall-runoff modelling is a cross cutting topic over several of the major issue this subject is focusing on. In view of its central role, rainfall-runoff modelling is then treated separately in this web page.

Rainfall-runoff models describe a portion of the water cycle and therefore the movement of a fluid - water - and therefore they are explicitly or implicitly based on the laws of physics, and in particular on the principles of conservation of mass, conservation of energy and conservation of momentum. Depending on their complexity, models can also simulate the dynamics of water quality, ecosystems, and other dynamical systems related to water, therefore embedding laws of chemistry, ecology, social sciences and so forth.

Models are built by constitutive equations, namely, mathematical formulations of the above laws, whose number depends on the number of variables to be simulated. The latter are the output variables, and the state variables, which one may need to, introduce to describe the state of the system. Constitutive equations may include parameters: they are numeric factors in the model equations that can assume different values therefore making the model flexible. In order to apply the model, parameters needs to be estimated (or calibrated, or optimized, and we say that the model is calibrated, parameterized, optimized). Parameters usually assume fixed value, but in some models they may depend on time, or the state of the system.

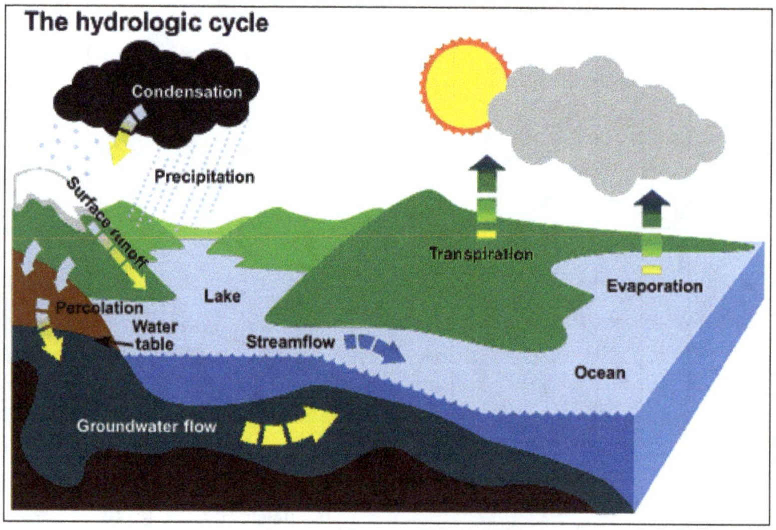

Figure: The hydrologic cycle.

Rainfall-runoff models can be classified within several different categories. They can distinguished between event-based and continuous-simulation models, black-box versus conceptual versus process based (or physically based) models, lumped versus distributed models, and several others. It is important to note that the above classifications are not rigid - sometimes a model cannot be unequivocally assigned to one category. We will treat rainfall-runoff models by taking into consideration models of increasing complexity.

The Rational Formula

The Rational equation is the simplest method to determine peak discharge from drainage basin runoff. It but is the most common method used for sizing sewer systems.

Rational Equation: Q=ciA

The Rational equation requires the following units:

Q = Peak discharge, cfs
c = Rational method runoff coefficient
i = Rainfall intensity, inch/hour
A = Drainage area, acre

Note that our calculation allows you to use a variety of units.

The Rational method runoff coefficient (c) is a function of the soil type and drainage basin slope.

The Linear Reservoir

The linear reservoir is one of the most used rainfall-runoff models, together with the time area method. The linear reservoir model assimilates the catchment to a reservoir, for which the conservation of mass applies. The reservoir is fed by rainfall, and releases the river flow through a bottom discharge, for which a linear dependence applies between the river flow and the volume of water stored in the reservoir, while other losses - including evapotranspiration - are neglected. Therefore, the model is constituted by the following relationships,

$$\frac{dW(t)}{dt} = p(t) - q(t)$$

and

$$W(t) = k\,q(t)$$

where W(t) is the volume of water stored in the catchment at time t, p(t) is rainfall, q(t) is the river flow at time t and k is a constant parameter with the dimension of time (if the parameter was not constant the model would not be linear).

The second equation above assumes a linear relationship between discharge and storage into the catchment. The properties of a linear function are described here. Actually, the relationship between storage in a real tank and bottom discharge is an energy conservation equation that is not linear; in fact, as it is given by the well-known Torricelli's law. Therefore the linearity assumption is an approximation, which is equivalent to assuming that the superposition principle applies to runoff generation. Actually, such assumption does not hold in practice, as the catchment response induced by two subsequent rainfall events cannot be considered equivalent to the sum of the individual catchment responses to each single event. However, linearity is a convenient assumption to make the model simpler and analytical integration possible.

A nice feature of the linear reservoir is that the above equations can be integrated analytically, under simplifying assumption. In fact, by substituting the second equation into the first one gets:

$$k\frac{dq(t)}{dt} = p(t) - q(t)$$

Then, by multiplying both sides by $e^{t/k}$ and dividing by k one gets:

$$e^{t/k}\frac{dq(t)}{dt}+e^{t/k}\frac{q(t)}{k}-e^{t/k}\frac{p(t)}{k}=0$$

which can be written as:

$$\frac{d}{dt}\left[e^{t/k}q(t)\right]=e^{t/k}\frac{p(t)}{k}$$

By integrating between 0 and t one obtains:

$$\int_0^t \frac{d}{d\tau}\left[e^{\tau/k}q(\tau)\right]d\tau=\int_0^t e^{\tau/k}\frac{p(\tau)}{k}d\tau$$

$$\left[e^{\tau/k}q(\tau)\right]_0^t=\int_0^t e^{\tau/k}\frac{p(\tau)}{k}d\tau$$

$$q(t)e^{t/k}-q(0)=\int_0^t e^{\tau/k}\frac{p(\tau)}{k}d\tau$$

and, by assuming q(0)=0, one finally gets:

$$q(t)=\int_0^t \frac{1}{k}e^{\frac{-(t-\tau)}{k}}p(\tau)d\tau$$

which can be easily integrated numerically by using the Euler method.

If one assumes p(t)=constant, an explicit expression is readily obtained for the river flow:

$$q(t)=\frac{p}{k}\int_0^t e^{-\frac{t-\tau}{k}}d\tau=\frac{p}{k}e^{-\frac{t}{k}}\int_0^t e^{\frac{\tau}{k}}d\tau=\frac{p}{k}e^{-\frac{t}{k}}\left[ke^{-\frac{t}{k}}\right]_0^t=\frac{p}{k}e^{-\frac{t}{k}}\left[ke^{\frac{t}{k}}-k\right]=p\left[1-e^{-\frac{t}{k}}\right]$$

The parameter k is usually calibrated by matching observed and simulated river flows. Its value significantly impacts the catchment response. A high k value implies a large storage into the catchment. Therefore, large values of k are appropriate for catchment with a significant storage capacity. Conversely, a low k value is appropriate for impervious basins. k is also related to the response time of the catchment. A low k implies a quick response, while slowly responding basins are characterized by a large k. In fact, k is related to the response time of the catchment.

The non-linear Reservoir

Several variants of the linear reservoir modeling scheme can be introduced, for instance by adopting a non linear relationship between discharge and storage. Moreover, an upper limit can be fixed for the storage in the catchment, and additional discharges can be introduced, which can be activated for different levels of storage. All of the above modifications make the model non-linear so that an analytical integration is generally not possible. One should also take into account that increasing the number of parameters implies a corresponding increase of estimation variance and therefore simulation uncertainty. Furthermore, it is often observed that introducing additional discharges or thresholds may induce discontinuities in the hydrograph shape.

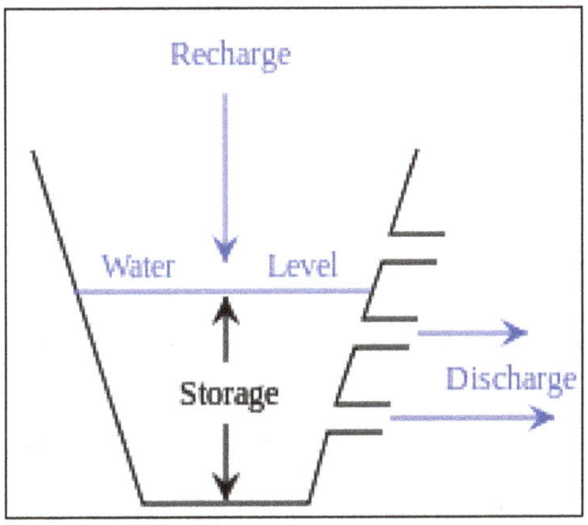

Figure: A non-linear reservoir.

The Hymod Model

The Hymod model is a flexible solution that is increasingly adopted for its capability of providing a good fit in several practical applications. It was originally proposed by Boyle (2000). It is based on the assumption that each point location i in the basin is characterised by a local value of soil water storage Ci, which varies from 0 in the impervious areas up to a maximum value Cmax in the most permeable location of the catchment. Ci is assumed to be randomly varying, so that for an assigned value C* of soil water storage a probability distribution is introduced that gives the probability that a randomly selected location j is characterised by Cj less then, or equal to, C*. Such probability may be interpreted as the fraction F(C*) of the catchment area where Cj≤C*. The above probability distribution is written as:

$$F(c_*) = 1 - \left(1 - \frac{c_*}{c_{\max}}\right)^{\beta k}, \quad o \le c_* \le c_{\max}$$

Here, βk is a parameter which quantifies the variability of the soil water storage over the catchment. One can easily verify by numerical simulation that βk=0 implies that the soil water storage is constant over the basin and equal to $Cmax$; $\beta k = 1$ implies that the soil water storage is linearly varying from 0 to Cmax; βk→∞ implies that the soil water storage is tending to the null value over the whole catchment, which is therefore impervious.

Let us assume that a storm event occurs over the basin and let us define with the symbol C(t) the time varying water depth stored in the unsaturated locations of the catchment. If we ignore any water losses, like evapotranspiration, C(t) is equal to the rainfall amount from the beginning of the event. If one assumes that the shape of the above probability distribution, now expressed in terms of C(t), is the one reported in figure below, it can be easily proved that the water volume stored in the catchment at time t is given by:

$$w(t)=c(t)-\left[\int_0^{c(t)} f(\chi)d\chi\right]$$

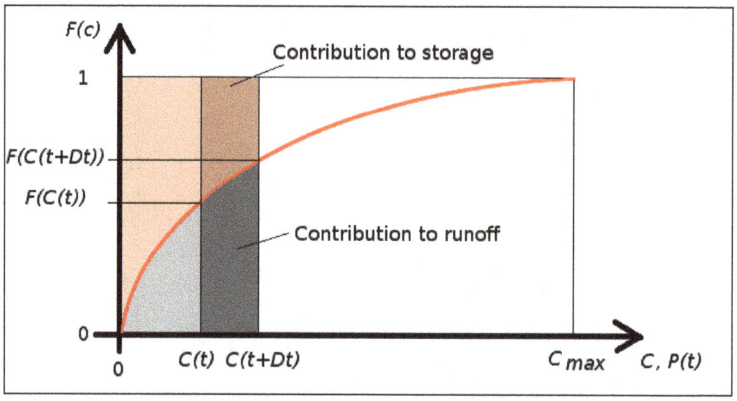

Figure: Distribution of soil water storage, surface runoff and stored volume in Hymod.

In fact, the integral at the right hand side of the above equation is the area below the red line in figure above. Elementary increment of that area are given by the product of the rainfall at each time step by the fraction F(c(t)) of saturated area at the same time, namely, the product of rainfall by saturated area, which is indeed the surface runoff. Conversely, the area above the curve gives the global storage into the catchment W(t), which can be interpreted as a weighted average value of C(t). The progress of surface runoff and water storage is depicted by the animated picture in figure below. Note that after saturation, the contribution of surface runoff is given by the product of the rainfall itself by 1, which is the fraction of saturated area, taking and keeping unit value when the catchment is saturated. After saturation, the storage in the catchment of course does not increase any more.

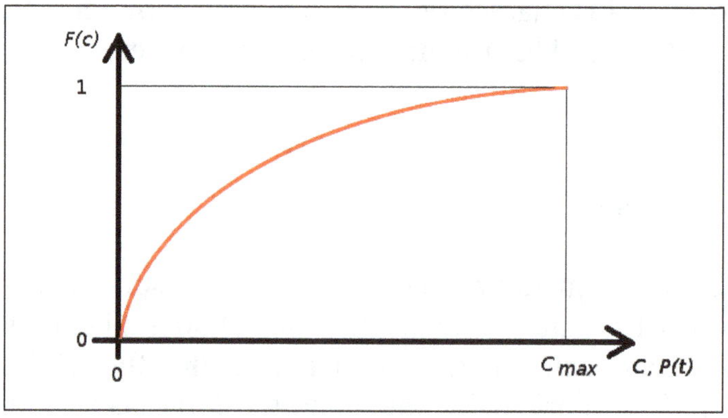

Figure: Progress of surface runoff and stored volume in Hymod.

By computing the integral in the latter relationship one obtains:

$$\int_0^{C(t)} F(\chi)dx=\int_0^{C(t)}\left\{1-\left[1-\frac{\chi}{C_{\max}}\right]^{\beta_k}\right\}d\chi$$

$$\int_0^{C(t)} F(\chi)dx = \left[\chi + \frac{C_{\max}}{(\beta_k+1)}\left[1-\frac{\chi}{C_{\max}}\right]^{\beta_k+1} \right]_0^{C(t)}$$

$$\int_0^{C(t)} F(\chi)d\chi = C(t) + \frac{C_{\max}}{(\beta_k+1)}\left[1-\frac{C(t)}{C_{\max}}\right]^{\beta_k+1} - \frac{C_{\max}}{(\beta_k+1)}$$

Therefore the water volume stored in the catchment is equal to,

$$W(t) = \frac{C_{\max}}{(\beta_k+1)} - \frac{C_{\max}}{(\beta_k+1)}\left[1-\frac{C(t)}{C_{\max}}\right]^{\beta_k+1} = \frac{C_{\max}}{(\beta_k+1)}\left\{1\left[1-\frac{C(t)}{C_{\max}}\right]^{\beta_k+1}\right\}$$

By inverting the above equation one gets:

$$C(t) = C_{\max}\left[1-\left(1-w(t)\frac{\beta_k+1}{c_{\max}}\right)^{\frac{1}{\beta_k+1}}\right]$$

Note that one obtains an estimate of the upper value of the water volume that the catchment can store by imposing C(t)=Cmax, therefore obtaining,

$$W_{\max} = \frac{C_{\max}}{(\beta_k+1)}$$

The above equations allow an easy application of the Hymod model through a numerical simulation, that is usually carried out by adopting a time step Dt that is equal to observational time step of rainfall and river flow. At a given time step t, one knows the value of C(t) which is equal to the cumulative rainfall depth from the beginning of the event at time t. Therefore, W(t) can be easily computed as well by using the above relationships. At the time t+1, C(t+1)=C(t)+P(t), where P is rainfall, under the condition that C(t+1)=Cmax if C(t)+P(t)>Cmax. Therefore, one can compute a first contribution to surface runoff through the relationship,

$ER_1(t)=\max(C(t)+P(t)-Cmax,0)$.

Finally, one can compute a second contribution to the surface runoff which is given by the water volume that cannot be absorbed by the catchment because part of the catchment area got saturated in the last time step. Such second contribution is given by,

$ER_2(t)=(C(t+1)-C(t))-(W(t+1)-W(t))$.

By summing ER1+ER2 one obtains the total contribution to the surface runoff during the time step from t to t+1.

At this stage, one may evaluate the water losses given by evapotranspiration at the current time step, and subtract them from W(t+1). Such water losses are computed within Hymod through the relationship,

$$E(t) = \left(1-\frac{\frac{C_{\max}}{\beta_k+1}-W(t)}{\frac{C_{\max}}{\beta_k+1}}\right)E_p(t)$$

Here, Ep(t) is the potential evapotranspiration at time t. Then, the water storage at time t+1 is given by,

$$W(t+1)=W(t)-Ep(t).$$

One should note that the evapotranspiration is subtracted from the stored water volume after ER1 and ER2 are computed.

The total contribution ER(t)=ER1(t)+ER2(t) to the surface runoff is then divided into 2 components: αER(t) which represent the fast runoff and (1-α)ER(t) which is the slow runoff. αER(t) is propagated through a series of linear reservoirs with the same bottom discharge time constant kq, while (1-α)ER(t) is instead propagated through a single linear reservoir with parameter ks.

The computation moves forward through the sequence of time steps. Hymod counts 5 parameters, namely: Cmax, β, α, kq and ks. These parameters need to be calibrated by using observed data.

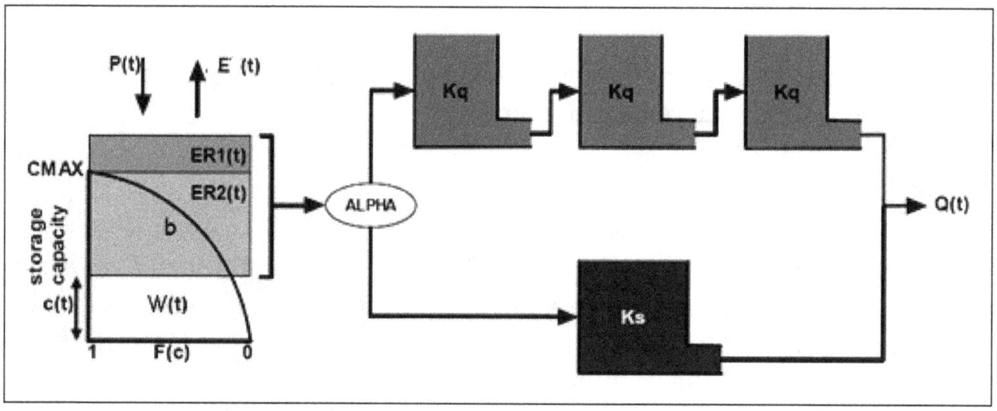

Figure: A schematic representation of the Hymod model.

Stormwater Runoff

Stormwater runoff occurs after a rainfall. Storm water flows over impervious (unable to penetrate) surfaces like driveways, sidewalks, streets, parking lots and roofs and is unable to percolate (filter or seep) into the ground. This unfiltered water reaches our neighborhood streams, ponds, lakes, bays, wetlands and oceans and can eventually make its way into our ground water.

Affect of Pollution on Watershed

There are two basic types of pollution: point source and non-point source. Point source pollution is easy to understand because it can be traced directly to its source. Point source pollution was a big concern in the past, but today stricter laws and regulations have drastically decreased the problem.

Non-point source pollution is a little more difficult to understand. Stormwater runoff pollution is a type of non-point source pollution. This means that the pollution cannot be traced back to a specific source, but instead comes from many different sources throughout the environment. Non-point source pollution is the primary cause of watershed pollution today. Non-point source pollution occurs when small amount of pollution from a large variety of sources is picked up by stormwater runoff and carried into water bodies.

Land cover affects.

Stormwater runoff can carry many different types of non-point source pollution. Each can affect your watershed in a different way. Sediment (dirt, soil, sand) can increase the turbidity (a measure of water cloudiness) of a water body. Turbidity can block sunlight from reaching aquatic plants, making it impossible for them to grow. Without plants, animals lose a food source and it is more difficult to filter pollutants from the water. Instead, pollutants collect in the bottom of the water body and remain there indefinitely.

Excess nutrients carried in stormwater runoff can also negatively affect our water supply. These nutrients, primarily nitrogen and phosphorus, can come from lawn fertilizers or natural sources, such as manure. Nutrients can cause algal and bacterial blooms, which proliferate (reproduce) rapidly. Algae will consume oxygen, increase turbidity in the water body and eventually die along with the fish and other aquatic life that need oxygen to live.

Debris such as plastic bags, bottles and cigarette butts can wash into a water body and interfere with aquatic life. Other hazardous wastes can be carried into a water body. These include insecticides, (chemicals used to control or kill insects) herbicides, (chemicals used to kill unwanted plants) paint, motor oil and heavy metals.

Stormwater Runoff: A Problem

Stormwater washes pollutants off roads, lawns and other surfaces and carries them into the nearest body of water. Some of the pollutants commonly carried by runoff are:

- Sediment
- Oil, grease, and toxic chemicals from motor vehicles
- Pesticides and nutrients from lawns and gardens
- Viruses, bacteria, and nutrients from pet waste and failing septic systems
- Heavy metals from roof shingles, motor vehicles, and other sources
- Road salts

These pollutants can harm fish and wildlife populations, kill native vegetation, foul drinking water supplies, and make recreational areas unsafe and unpleasant.

Stream Damage

Increased flow resulting from excess runoff causes extensive damage to streams, scouring the stream bottom and causing severe erosion to stream banks. The risk of flooding downstream also increases.

Stream health.

Urban Runoff

Urban runoff is either wet weather (rainwater) or dry weather (water waste) flows from urban landscapes into storm drain systems that lead to the beach.

Urban Runoff: A Problem

Urban runoff carries contaminants, such as litter, food, human & animal waste, automobile fluids, industrial pollutants, fertilizers and pesticides to the beach creating health risks for people, killing marine life and contributing to localized flooding and beach closures.

We also see the impacts of pollution in increased health risks to swimmers near flowing storm drains and toxicity to aquatic life. These impacts translate into losses to the tourism economy, loss of recreational resources, dramatic cost increases for cleaning up contaminated sediments, degraded water quality and impaired function and vitality of our natural resources.

Clogged storm drains can lead to area flooding when it rains, creating traffic problems and unsanitary conditions.

Urban Flooding

Flooded streets in New Orleans.

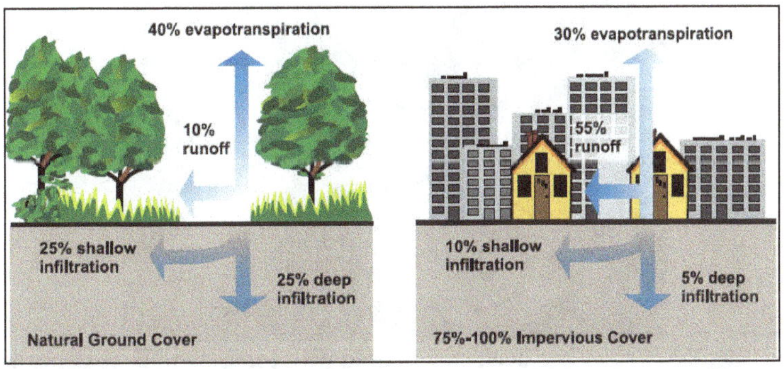

Relationship between impervious surfaces and surface runoff.

Urban runoff is a major cause of urban flooding, the inundation of land or property in a built-up environment caused by rainfall overwhelming the capacity of drainage systems, such as storm sewers. Triggered by events such as flash flooding, storm surges, overbank flooding, or snow melt, urban flooding is characterized by its repetitive, costly and systemic impacts on communities, regardless of whether or not these communities are located within formally designated floodplains or near any body of water.

There are several ways in which stormwater enters properties: backup through sewer pipes, toilets and sinks into buildings; seepage through building walls and floors; the accumulation of water on property and in public rights-of-way; and the overflow of water from water bodies such as rivers and lakes. Where properties are built with basements, urban flooding is the primary cause of basement flooding.

Flood flows in urban environments have been investigated relatively recently despite many centuries of flood events. Some researchers mentioned the storage effect in urban areas. Several studies looked into the flow patterns and redistribution in streets during storm events and the implication in terms of flood modelling. Some recent research considered the criteria for safe evacuation of individuals in flooded areas. But some recent field measurements during the 2010–2011 Queensland floods showed that any criterion solely based upon the flow velocity, water depth or specific momentum cannot account for the hazards caused by the velocity and water depth fluctuations. These considerations ignore further the risks associated with large debris entrained by the flow motion.

Effects

A 2008 report by the United States National Research Council identified urban runoff as a leading source of water quality problems.

> ...further declines in water quality remain likely if the land-use changes that typify more diffuse sources of pollution are not addressed. These include land-disturbing agricultural, silvicultural, urban, industrial, and construction activities from which hard-to-monitor pollutants emerge during wet-weather events. Pollution from these landscapes has been almost universally acknowledged as the most pressing challenge to the restoration of waterbodies and aquatic ecosystems nationwide.
>
> – National Research Council, *Urban Stormwater Management in the United States*

Weasel Brook in Passaic, New Jersey has been channelized with concrete walls to control localized flooding.

The runoff also increases temperatures in streams, harming fish and other organisms. (A sudden burst of runoff from a rainstorm can cause a fish-killing shock of hot water.) Also, road salt used to melt snow on sidewalks and roadways can contaminate streams and groundwater aquifers.

One of the most pronounced effects of urban runoff is on watercourses that historically contained little or no water during dry weather periods (often called *ephemeral streams*). When an area around such a stream is urbanized, the resultant runoff creates an unnatural year-round streamflow that hurts the vegetation, wildlife and stream bed of the waterway. Containing little or no sediment relative to the historic ratio of sediment to water, urban runoff rushes down the stream channel, ruining natural features such as meanders and sandbars, and creates severe erosion—increasing sediment loads at the mouth while severely incising the stream bed upstream. As an example, on many Southern California beaches at the mouth of a waterway, urban runoff carries trash, pollutants, excessive silt, and other wastes, and can pose moderate to severe health hazards.

Because of fertilizer and organic waste that urban runoff often carries, eutrophication often occurs in waterways affected by this type of runoff. After heavy rains, organic matter in the waterway is relatively high compared with natural levels, spurring growth of algae blooms that soon consume most of the oxygen. Once the naturally occurring oxygen in the water is depleted, the algae blooms die, and their decomposition causes further eutrophication. Algae blooms mostly occur in areas with still water, such as stream pools and the pools behind dams, weirs, and some drop structures. Eutrophication usually comes with deadly consequences for fish and other aquatic organisms.

Excessive stream bank erosion may cause flooding and property damage. For many years governments have often responded to urban stream erosion problems by modifying the streams through construction of hardened embankments and similar control structures using concrete and masonry materials. Use of these hard materials destroys habitat for fish and other animals. Such a project may stabilize the immediate area where flood damage occurred, but often it simply shifts the problem to an upstream or downstream segment of the stream.

Urban flooding has significant economic implications. In the US, industry experts estimate that wet basements can lower property values by 10%-25% and are cited among the top reasons for not purchasing a home. According to the U.S Federal Emergency Management Agency (FEMA), almost 40% of small businesses never reopen their doors following a flooding disaster. In the UK, urban flooding is estimated to cost £270 million a year in England and Wales; 80,000 homes are at risk.

A study of Cook County, Illinois, identified 177,000 property damage insurance claims made across 96% of the county's ZIP codes over a five-year period from 2007-2011. This is the equivalent of one in six properties in the County making a claim. Average payouts per claim were $3,733 across all types of claims, with total claims amounting to $660 million over the five years examined.

Despite concerted efforts, many communities lack the funds to fully address these issues, and often seek funds elsewhere. Numerous watersheds within Los Angeles County, California do not meet state water quality standards, despite spending $100 million a year on clean water programs to combat issues such as urban runoff. To combat this problem, officials have introduced a measure that would assess a fee to homeowners and local businesses in attempt to raise $290 million for effective urban runoff management.

Prevention and mitigation

Effective control of urban runoff involves reducing the velocity and flow of stormwater, as well as reducing pollutant discharges. A variety of stormwater management practices and systems may be used to reduce the effects of urban runoff. Some of these techniques (called best management practices (BMPs) in the US), focus on water quantity control, while others focus on improving water quality, and some perform both functions.

A percolation trench infiltrates stormwater through permeable soils into the groundwater aquifer.

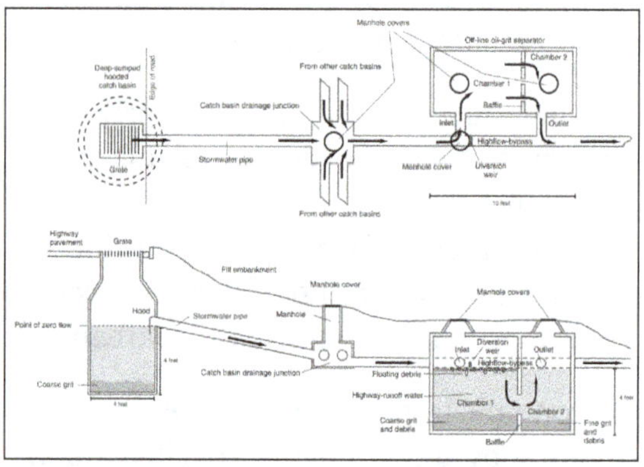

An oil-grit separator is designed to capture settleable solids, oil and grease, debris and floatables in runoff from roads and parking lots.

Pollution prevention practices include low impact development (LID) or green infrastructure techniques - known as Sustainable Drainage Systems (SuDS) in the UK, and Water-Sensitive Urban Design (WSUD) in Australia and the Middle East - such as the installation of green roofs and improved chemical handling (e.g. management of motor fuels & oil, fertilizers and pesticides). Runoff mitigation systems include infiltration basins, bioretention systems, constructed wetlands, retention basins and similar devices.

Providing effective urban runoff solutions often requires proper city programs that take into account the needs and differences of the community. Factors such as a city's mean temperature, precipitation levels, geographical location, and airborne pollutant levels can all effect rates of pollution in urban runoff and present unique challenges for management. Human factors such as urbanization rates, land use trends, and chosen building materials for impervious surfaces often exacerbate these issues.

The implementation of citywide maintenance strategies such as street sweeping programs can also be an effective method in improving the quality of urban runoff. Street sweeping vacuums collect particles of dust and suspended solids often found in public parking lots and roads that often end up in runoff.

Educational programs can also be an effective tool for managing urban runoff. Local businesses and individuals can have an integral role in reducing pollution in urban runoff simply through their practices, but often are unaware of regulations. Creating a productive discussion on urban runoff and the importance of effective disposal of household items can help to encourage environmentally friendly practices at a reduced cost to the city and local economy.

Blue drain and yellow fish symbol used by the UK Environment Agency to raise awareness of the ecological impacts of contaminating surface drainage.

Runoff Model

Outflow at the bottom of the soilwater box, f, forms inflow to the runoff box, i.e it corresponds to peff, as it has been described in the section on reservoir models. A conceptual runoff model can therefore be illustrated as in figure below. One has to solve the continuity equations for soilwater eq. and for the runoff box eq. ($\frac{ds}{dt} = p - e - q$) and the accompanying runoff (the equations are given new numbering here):

$$\frac{d_{h_{mark}}}{d\,t} 2 = p - e - f; e, f = funk\left(h_{mark}\right)$$

$$\frac{dh}{d\,t} 3 = f - q; q = funk\left(h\right)$$

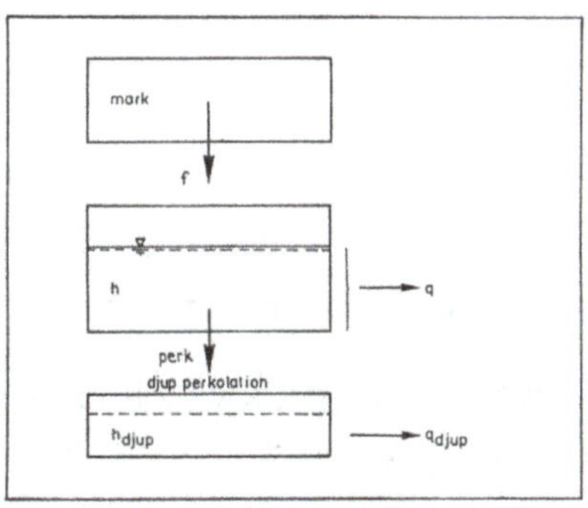

Figure: Runoff model with soilwater reservoir and double runoff reservoir.

The solution is most easily done with an explicit method according to the scheme given in figure above. hmark is related to hwilt, which is taken to be zero.

$t = 0$; värden på h_{mark}, h, e, f, q

→ $t = t + dt$; läs p och pe

→ för kort effektivt tidsteg utförs

$h_{mark} = h_{mark}$ (gamla) + $(p - e - f)\, dt_{eff}$

$h = h$ (gamla) + $(f - q)\, dt_{eff}$

f =; e =; q =;

till dess antal dt_{eff} motsvarar dt

p, q, e skrivs ut

ny tid om $t < t_{sluta}$

Figure: Explicit solution in runoff model.

The runoff model can be made more complex than shown above. In the model shown in figure above there are 5 parameters, hp; FC; To; T1; ho. The more parameters introduced the more difficult it is to adjust the model to real general situations even if special events can be modeled almost perfectly. One can include surface runoff from the soilwater box, one can use several holes, i.e. more time constants, in the runoff box. One can allow evaporation directly from parts of the runoff box and one can let some precipitation go straight to the runoff box. One can let runoff go into another box. There are several modifications all of which can be motivated in special situations.

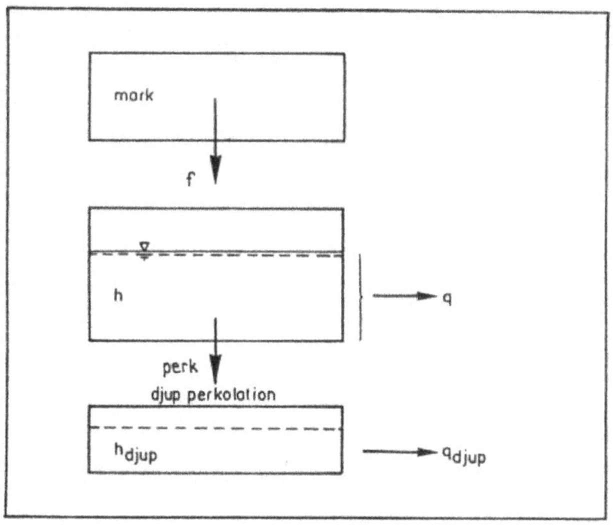

Figure: Runoff model with soilwater box , runoff reservoir and a deep water storage.

An addition to the runoff model shown in figure above which is very often used is a deep groundwater box, which is used to model slow groundwater flow forming base flow. Water is supposed to percolate from the runoff box into the deep groundwater box, as shown in figure above. In the Swedish HBV-model smaller lakes are taken as part of the groundwater storage in the sense that precipitation on the lake surface goes directly to deep storage and evaporation from the lake surface constitutes a loss from the deep storage. The rate of percolation, perk, from groundwater box to deep groundwater-box can be constant or depend on amount of stored water. The continuity equations for the runoff box and the deep groundwater box becomes respectively:

$$\frac{dh}{dt} = f - q - perk \quad 4$$

$$\frac{d\,h_{djup}}{d\,t} = perk - q_{djup} \quad 5$$

$$qd_{djup} = \frac{h_{djup}}{T_{djup}} \quad 6$$

The runoff is the summation of q and d_{djup}

Before one has a complete runoff model the three boxes described here for soilwater, groundwater and deep groundwater must be supplemented with interception- and snow models for calculation of inflow to the soilwater box, and with a routing model which describes how the runoff is delayed in the system of streams and rivers in the catchment.

Runoff Curve Number Method

The SCS Runoff Curve Number method is developed by the United States Department of Agriculture (USDA) Soil Conservation Service (SCS) and is a method of estimating rainfall excess from rainfall.

The conceptual basis of the curve number method has been the object of both support and criticism (Ponce and Hawkins, 1996). The major disadvantages of the method are sensitivity of the method to Curve Number (CN) values, fixing the initial abstraction ratio, and lack of clear guidance on how to vary Antecedent Moisture Conditions (AMC). However, the method is used widely and is accepted in numerous hydrologic studies. The SCS method originally was developed for agricultural watersheds in the mid-western United States; however it has been used throughout the world far beyond its original developers would have imagined.

The basis of the curve number method is the empirical relationship between the retention (rainfall not converted into runoff) and runoff properties of the watershed and the rainfall. Mockus found equation 1 appropriate to describe the curves of the field measured runoff and rainfall values (National Engineering Handbook, 2004). Equation 1 describes the conditions in which no initial abstraction occurs.

$$\frac{F}{S} = \frac{Q}{P}$$

where F = P – Q = actual retention after runoff begins;

Q = actual runoff

S = potential maximum retention after runoff begins (S ³ F)

P = potential maximum runoff (i.e., total rainfall if no initial abstraction).

For most applications, a certain amount of rainfall is abstracted. The three important abstractions for any single storm event are rainfall interception (Meteorological rainfall minus throughfall, stem flow and water drip), depression storage (topographic undulations), and infiltration into the soil. The curve number method lumps all three abstractions into one term, the Initial abstraction (Ia), and subtracts this calculated value from the rainfall total volume. The total rainfall must exceed this initial abstraction before any runoff is generated. This gives the potential maximum runoff (rainfall available for runoff) as P – Ia. Substituting this value in equation 1 yields following equation:

$$\frac{P-Ia-Q}{S}=\frac{Q}{P-Ia}$$

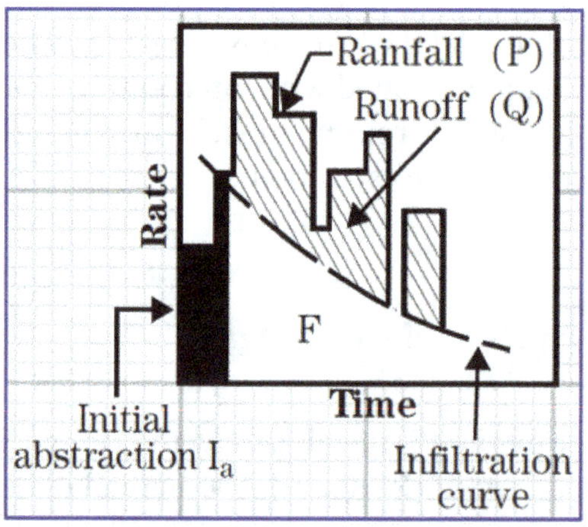

Figure: Components of SCS Runoff equation.

It is important to note the potential maximum retention term, "S", excludes Ia. Hence, for a given storm, maximum loss of rainfall is S plus Ia. Rearranging terms in Equation ($\frac{P-Ia-Q}{S}=\frac{Q}{P-Ia}$) for Q gives

$$Q=\frac{(P-Ia)^2}{(P-Ia)+S}$$

Establishing the relation to estimate Ia was challenging. The SCS provided the following empirical equation below based on the assumption Ia was a function of the potential maximum retention S.

Ia = 0.2S

The potential maximum retention S is related to the dimensionless parameter CN in the range of 0 <= CN <= 100 by equation below.

$$S=\left(\frac{1000}{CN}\right)-10)$$

Substituting Equation (Ia = 0.2S) into Equation ($Q = \frac{(P - Ia)^2}{(P - Ia) + S}$) yields,

$$Q = \frac{(P - 0.2S)^2}{(P + 0.8S)}$$

Equation above has only one parameter that needs to be evaluated (i.e., S) which can be determined by using equation ($S = \left(\frac{1000}{CN}\right) - 10)$) and curve number tables published by the SCS.

The solution of the SCS runoff equation is shown below.

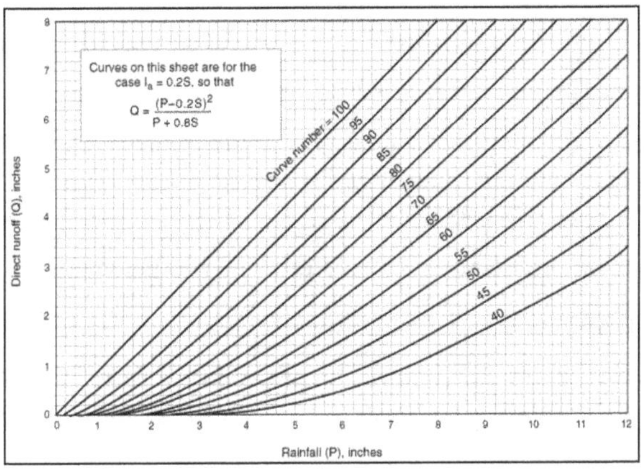

When the drainage area consists of land use with impervious cover (Directly Connected or Unconnected), TR 55 provides separate graphs for computing the composite curve number values depending on the percent of the impervious cover. However, it is a good practice to calculate the runoff from impervious area and pervious area separately and then add the volume.

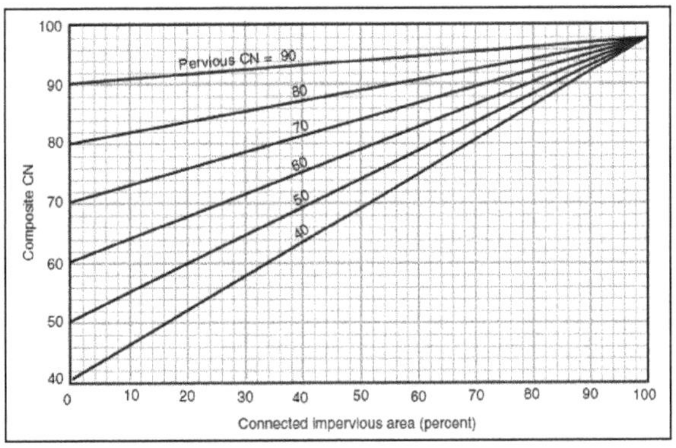

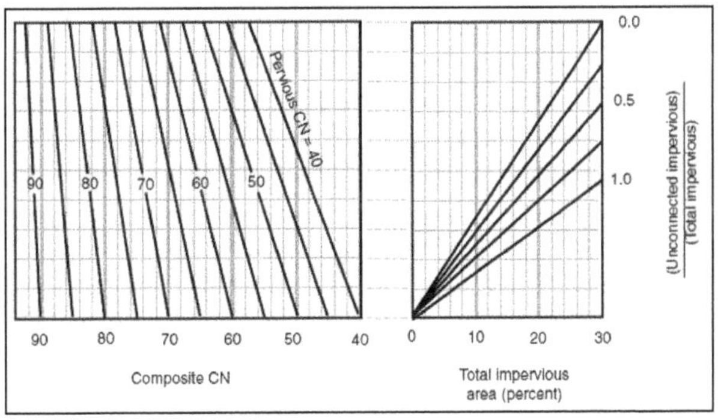

References

- “Parking Lot and Street Cleaning”. National Menu of Stormwater Best Management Practices. EPA. Archived from the original on 2015-08-28. Retrieved 05, March 2020
- What-is-snowmelt: albertawater.com, Retrieved 10, May 2020
- Ballo, Siaka; Liu, Min; Hou, Lijun; Chang, Jing (2009-07-10). “Pollutants in stormwater runoff in Shanghai (China): Implications for management of urban runoff pollution”. Progress in Natural Science. 19 (7): 873–880. doi:10.1016/j.pnsc.2008.07.021
- Fundamentals: physicalgeography.net, Retrieved 19, February 2020
- National Research Council (United States) (2009). Urban Stormwater Management in the United States (Report). Washington, D.C.: National Academies Press. p. 24. doi:10.17226/12465. ISBN 978-0-309-12539-0
- Watercyclesnowmelt: water.usgs.gov, Retrieved 06, June 2020
- Sansalone, John; Christina, Chad M. (November 2004). “First flush concepts for dissolved solids in small impervious watersheds”. Journal of Environmental Engineering. 130 (11): 1301–1314. doi:10.1061/(ASCE)0733-9372(2004)130:11(1301). ISSN 0733-9372
- Curve-number-introduction: professorpatel.com, Retrieved 11, August 2020
- Stormwater: rivanna-stormwater.org, Retrieved 01, July 2020
- Werner, MGF; Hunter, NM; Bates, PD (2006). “Identifiability of Distributed Floodplain Roughness Values in Flood Extent Estimation”. Journal of Hydrology. 314: 139–157. doi:10.1016/j.jhydrol.2005.03.012
- Stormwater-education: villagehohny.org, Retrieved 28, January 2020

6

A Statistical Approach to Hydrology

Statistical evaluation in hydrology includes the statistical procedures for evaluating the daily and monthly predictions of hydrologic models. This chapter elaborates statistical analysis, probability distribution and statistical hypothesis testing. All the diverse principles of statistical evaluation have been carefully analyzed in this chapter.

Statistical Analysis

A hydrologic process is a phenomenon describing the occurrence and movement of water in the earth phase of the hydrologic cycle. When the phenomena are observed at intervals of time, it is termed as a discrete process and when these are observed continuously, continuous series are generated. The observed behavior of many hydrologic variables cannot be completely explained by applying the physical laws. Three reasons behind this are: a) incorrect or incomplete knowledge of the underlying processes, b) inadequate data, and c) inherent randomness of the process. In these situations we resort to statistical analysis to infer and predict the behavior of the hydrologic variables.

Data of the concerned process are collected to apply the probability theory. All possible observations of a process constitute its population. But we cannot completely collect the data of population in a limited time frame. Generally, only a finite portion of the population is observed and is called a sample. Statistical properties of the sample are determined and it is assumed that these represent the properties of the population.

Many hydrologic data (e.g., daily rainfall amount, discharge of a river) can be characterized by variable(s) that are unpredictable to some degree. Yet frequently there is a degree of consistency in the factors governing the outcome which exhibits a statistical regularity. A variable whose value at any time is not influenced by the value at earlier time(s) is known as a *random variable*. A random variable can be discrete which means that it can take on only a finite set of values, for example the number of rainy days in a year at Roorkee. It can also be continuous (a time series) and can take on any value, for example, the water level of Ganga River at Rishikesh or the magnitude of rainfall at a place.

The behavior of the hydrological processes may be deterministic or stochastic. For a deterministic process, the behavior can be completely described with certainty. The governing equation defines the process for the entire time (or space) domain. In contrast to this, a stochastic process evolves, entirely or in major part, according to a random mechanism and hence the future outcomes of the process are not exactly predictable. Such hydrologic variables whose values are governed by the laws of chance are called stochastic variables.

For many hydrological problems in these cases, sample data consist of measurements made on a single random variable and the techniques of analysis are called univariate analysis and estimation. Univariate analysis is carried out by using the measurements of the random variable (or the sample information) to identify the statistical properties of the population from which the sample measurements are likely to have come. After the underlying population has been identified, one can make probabilistic statements about the future occurrences of the random variable; this represents univariate estimation.

Before commencing the analysis, it is always useful to plot the sample data. The following are the main steps of statistical analysis of data:

i) Compute the basic statistical indicators of the data.

ii) Select a set of probability distribution functions.

iii) Fit the selected distributions with the sample data. Common methods of parameter estimation are discussed subsequently.

iv) Select the best fit distribution using the goodness-of-fit tests.

v) Use the best-fit probability distribution to make inferences about the likelihood of occurrence of the magnitudes of the random variable.

If all the values of a random variable and the corresponding probabilities are known or found, the relation between these values and probabilities is described by a probability distribution. Knowing this distribution, the probability of any value of the random variable can be determined.

Probability

Based on the daily experiences, most people have an intuitive appreciation of the concept of probability or chance. When a coin is thrown, there are two equally likely possible outcomes: head or tail. Let the coin be tossed n times and head occurs s times. The ratio s/n is the probability of occurrence of heads.

Sample space is a collection of all possible random events that might arise from a random experiment. Sample space S and two sets of events A and B are shown in figure.

If two events A and B do not contain any common sample points, they are said to be mutually exclusive.

If two events A and B are not mutually exclusive, the common set is called their *intersection*, denoted as $A \cap B$. The union of two events A and B gives the event which is the collection of all sample points occurring at least once in either A or B and is denoted as $A \cup B$. Figure shows this concept through the Venn diagram. The complement A^c of an event A consists of all sample points in the sample space of the experiment not included in the event A.

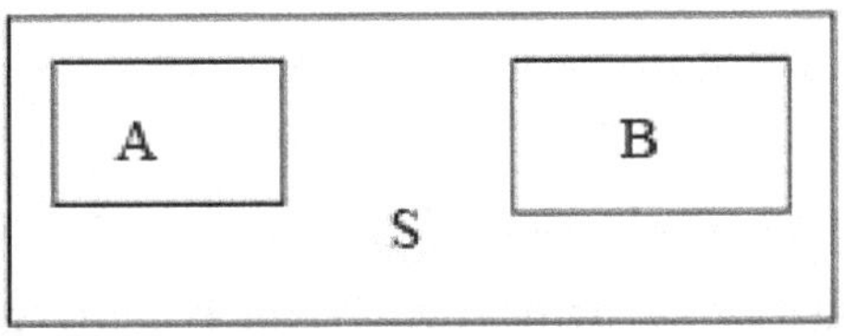

Two mutually exclusive events *A* and *B* in sample space *S*.

$A \cap B$ $A \cup B$

Venn diagram showing intersection and union of two events.

Axioms of Probabilities

Events are basically sets and the common set operators, including union, intersection, and complement, are applicable to them. Four frequently used rules for set operations, considering three events A, B and C are:

1. Commutative rule: $A \cup B = B \cup A$, $A \cap B = B \cap A$
2. Associative rule: $(A \cup B) \cup C = A \cup (B \cup C)$, $(A \cap B) \cap C = A \cap (B \cap C)$
3. Distributive rule: $A \cap (B \cup C) = (A \cap B) \cup (A \cap C)$, $A \cup (B \cap C) = (A \cup B) \cap (A \cup C)$
4. De Morgan's rule: $(A \cup B)^C = A^C \cap B^C$, $(A \cup B)^C = A^C \cap B^C$

The notation $P[A]$ is used to denote the probability of a random event *A*. Now we discuss the axioms of probability.

Axiom 1: The probability of an event *A* is a number greater than or equal to zero but less than or equal to unity:

$$0 \le P[A] \le 1$$

Axiom 2: The probability of an event A, whose occurrence is certain, is unity:

$$P[A] = 1$$

where A is the event associated with all sample points in the sample space.

Axiom 3: The probability of an event which is the union of two events is:

$$P[A \text{ or } B] = P[A \cup B] = P[A] + P[B] - P[A \cap B]$$

where $A \cup B$ denotes the union of events A and B which means that either event A occurs or event B occurs, and $A \cap B$ denotes the intersection of event A and event B. Eq. can be extended to the union of n events. If A and B are two mutually exclusive (disjointed) events, the probability of A and B, $P[A \cap B]$, will be zero and eq. becomes:

$$P[A \text{ or } B] = P[A] + P[B]$$

Axiom 4: The probability of two (statistically) independent events occurring simultaneously or in succession is the product of individual probabilities:

$$P[E_1 \cap E_2] = P[E_1] \times P[E_2]$$

Statistical independence implies that the occurrence of event E_1 has no influence on the occurrence of event E_2.

Properties of Random Variable

Let X denote a random variable and x be a possible value of X. The cumulative distribution function (CDF), $F_X(x)$ is the probability that the random variable X is less than or equal to x:

$$F_X(x) = \mathrm{P}(\mathrm{X} \le \mathrm{x})$$

The probability distribution function (PDF) describes the relative likelihood that a continuous random variable X takes on different values, and is the derivative of the CDF:

$$\mathrm{f_X(x) = d\{F_X(x)\} / dx}$$

The PDF and CDF of a random variable are shown in figure. Note that the CDF is denoted by capital letter F and the PDF by lower case letter f.

Let us now state some important properties of $f(x)$ and $F(x)$ for continuous and discrete random variables.

1. The probability of a random variable cannot be negative:

$$f(x) \geq 0, \quad -\infty < x < \infty$$

2. The sum of probabilities of all possible outcomes is equal to 1, i.e., the area under the PDF is unity.

$$\int_{-\infty}^{\infty} f(x)dx = 1$$

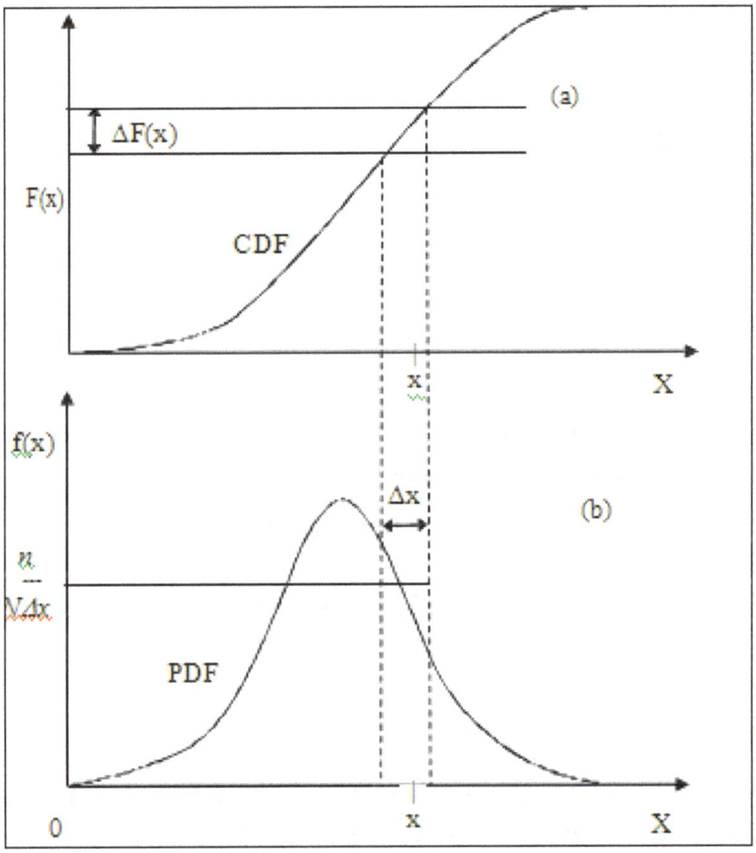

PDF and CDF of a random variable.

For discrete random variables, the following statements can be made:

$$\sum_i f(x) = 1$$

where $f(x_i)$ represents the probability of $X = x_i$ in the sample space if the sample contains finite and $f(x_i)$ observations can be replaced by $p(x_i)$. The probability of "X assumes a value $\leq$ x" is equal to the area under PDF curve to the left of x:

3. $P(X \leq x) = F(X \leq x) = F(x) = \int_{-\infty}^{x} f(x)dx$

Moreover

$$P(a < X \le b) = P(X \le b) - P(X \le a)$$
$$= \int_{-\infty}^{b} f(x)dx - \int_{-\infty}^{a} f(x)dx$$
$$= \int_{a}^{b} f(x)dx, \quad for\ a < b$$

For discrete case

$$P(a \le x \le b) = \sum_{x_i \ge a}^{x_i \le b} p(x_i)$$

$$P(X \le x_k) = \sum_{i=1}^{k} p(x_i)$$

Probability Distribution

In probability theory and statistics, a probability distribution is a mathematical function that, stated in simple terms, can be thought of as providing the probability of occurrence of different possible outcomes in an experiment. For instance, if the random variable X is used to denote the outcome of a coin toss ('the experiment'), then the probability distribution of X would take the value 0.5 for $X = \text{heads}$, and 0.5 for $X = \text{tails}$.

In more technical terms, the probability distribution is a description of a random phenomenon in terms of the probabilities of events. Examples of random phenomena can include the results of an experiment or survey. A probability distribution is defined in terms of an underlying sample space, which is the set of all possible outcomes of the random phenomenon being observed. The sample space may be the set of real numbers or a higher-dimensional vector space, or it may be a list of non-numerical values; for example, the sample space of a coin flip would be $\{\text{heads}, \text{tails}\}$.

Probability distributions are generally divided into two classes. A discrete probability distribution (applicable to the scenario where the set of possible outcomes is discrete, such as a coin toss or a roll of dice) can be encoded by a discrete list of the probabilities of the outcomes, known as a probability mass function. On the other hand, a continuous probability distribution (applicable to the scenarios where the set of possible outcomes can take on values in a continuous range (e.g., real numbers), such as the temperature on a given day) is typically described by probability density functions (with the probability of any individual outcome actually being 0). The normal distribution represents a commonly encountered continuous probability distribution. More complex experiments, such as those involving stochastic processes defined in continuous time, may demand the use of more general probability measures.

A probability distribution whose sample space is the set of real numbers is called univariate, while a distribution whose sample space is a vector space is called multivariate. A univariate distribution gives the probabilities of a single random variable taking on various alternative values; a multivariate distribution (a joint probability distribution) gives the probabilities of a random vector—a list of two or more random variables—taking on various combinations of values. Important and commonly encountered univariate probability distributions include the binomial distribution, the hypergeometric distribution, and the normal distribution. The multivariate normal distribution is a commonly encountered multivariate distribution.

Introduction

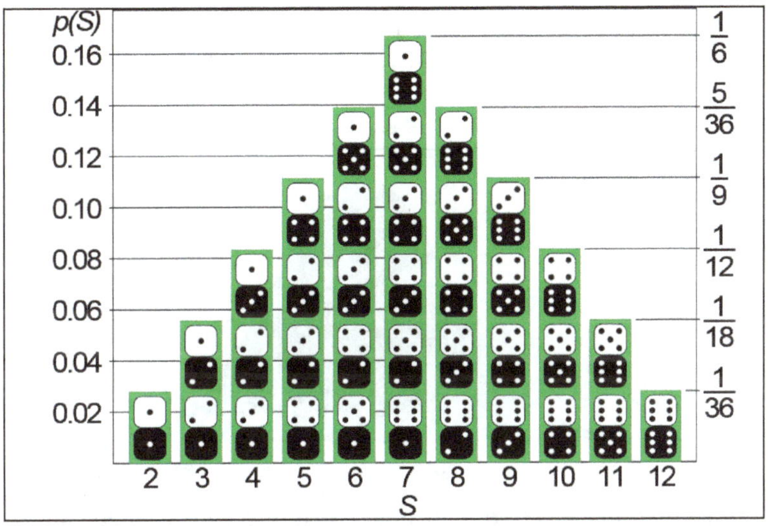

The probability mass function (pmf) $p(S)$ specifies the probability distribution for the sum S of counts from two dice. For example, the figure shows that $p(11) = 1/18$. The pmf allows the computation of probabilities of events such as $P(S > 9) = 1/12 + 1/18 + 1/36 = 1/6$, and all other probabilities in the distribution.

To define probability distributions for the simplest cases, one needs to distinguish between discrete and continuous random variables. In the discrete case, it is sufficient to specify a probability mass function assigning a probability to each possible outcome: for example, when throwing a fair dice, each of the six values *1* to *6* has the probability 1/6. The probability of an event is then defined to be the sum of the probabilities of the outcomes that satisfy the event; for example, the probability of the event "the die rolls an even value" is,

$$\text{Prob}(2) + \text{Prob}(4) + \text{Prob}(6) = 1/6 + 1/6 + 1/6 = 1/2.$$

In contrast, when a random variable takes values from a continuum then typically, any individual outcome has probability zero and only events that include infinitely many outcomes, such as intervals, can have positive probability. For example, the probability that a given object weighs *exactly* 500 g is zero, because the probability of measuring exactly 500 g tends to zero as the accuracy of our measuring instruments increases. Nevertheless, in quality control one might demand that the probability of a "500 g" package containing between 490 g and 510 g should be no less than 98%, and this demand is less sensitive to the accuracy of our instruments.

Continuous probability distributions can be described in several ways. The probability density function describes the infinitesimal probability of any given value, and the probability that the outcome lies in a given interval can be computed by integrating the probability density function over that interval. On the other hand, the cumulative distribution function describes the probability that the random variable is no larger than a given value; the probability that the outcome lies in a given interval can be computed by taking the difference between the values of the cumulative distribution function at the endpoints of the interval. The cumulative distribution function is the antiderivative of the probability density function provided that the latter function exists.

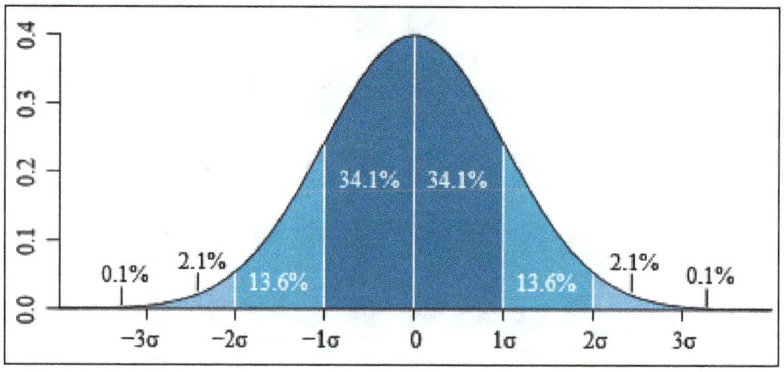

The probability density function (pdf) of the normal distribution, also called Gaussian or "bell curve", the most important continuous random distribution. As notated on the figure, the probabilities of intervals of values correspond to the area under the curve.

Terminology

As probability theory is used in quite diverse applications, terminology is not uniform and sometimes confusing. The following terms are used for non-cumulative probability distribution functions:

- Distribution, Frequency distribution: is a table that displays the frequency of various outcomes in a sample.
- Probability distribution: is a table that displays the probabilities of various outcomes in a sample. Could be called a "normalized frequency distribution table", where all occurrences of outcomes sum to 1.

- Distribution function: is a functional form of frequency distribution table.
- Probability distribution function: is a functional form of probability distribution table. Could be called a "normalized frequency distribution function", where area under the graph equals to 1.

Finally,

- Probability mass, Probability mass function, p.m.f., Discrete probability distribution function: for discrete random variables.
- Categorical distribution: for discrete random variables with a finite set of values.
- Probability density, Probability density function, p.d.f., Continuous probability distribution function: most often reserved for continuous random variables.

The following terms are somewhat ambiguous as they can refer to non-cumulative or cumulative distributions, depending on authors' preferences:

- Probability distribution function: continuous or discrete, non-cumulative or cumulative.
- Probability function: even more ambiguous, can mean any of the above or other things.

Basic Terms

- Mode: for a discrete random variable, the value with highest probability (the location at which the probability mass function has its peak); for a continuous random variable, the location at which the probability density function has its peak.
- Support: the smallest closed set whose complement has probability zero.
- Head: the range of values where the pmf or pdf is relatively high.
- Tail: the complement of the head within the support; the large set of values where the pmf or pdf is relatively low.
- Expected value or mean: the weighted average of the possible values, using their probabilities as their weights; or the continuous analog thereof.
- Median: the value such that the set of values less than the median has a probability of one-half.
- Variance: the second moment of the pmf or pdf about the mean; an important measure of the dispersion of the distribution.

- Standard deviation: the square root of the variance, and hence another measure of dispersion.
- Symmetry: a property of some distributions in which the portion of the distribution to the left of a specific value is a mirror image of the portion to its right.
- Skewness: a measure of the extent to which a pmf or pdf "leans" to one side of its mean. The third standardized moment of the distribution.
- Kurtosis: a measure of the "fatness" of the tails of a pmf or pdf. The fourth standardized moment of the distribution.

Cumulative Distribution Function

Because a probability distribution Pr on the real line is determined by the probability of a scalar random variable X being in a half-open interval $(-\infty, x]$, the probability distribution is completely characterized by its cumulative distribution function:

$$F(x) = \Pr[X \leq x] \qquad \text{for all } x \in \mathbb{R}.$$

Discrete Probability Distribution

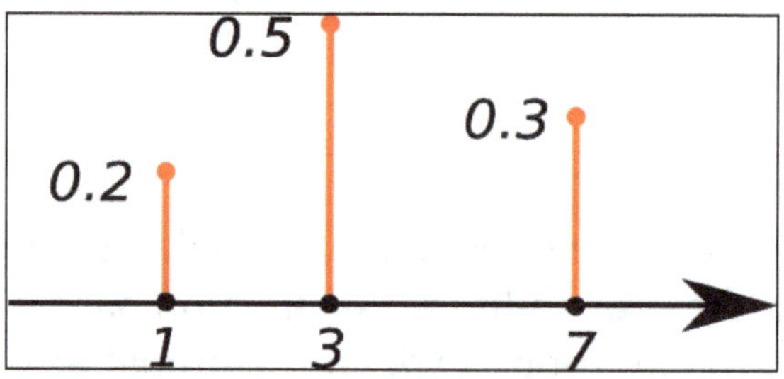

The probability mass function of a discrete probability distribution. The probabilities of the singletons {1}, {3}, and {7} are respectively 0.2, 0.5, 0.3. A set not containing any of these points has probability zero.

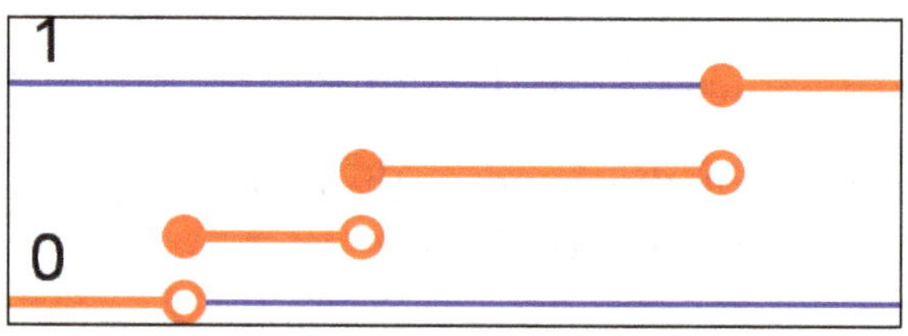

The cdf of a discrete probability distribution, ...

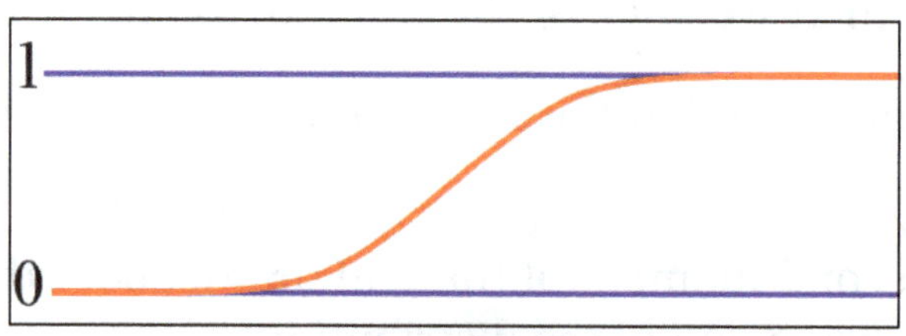

... of a continuous probability distribution, ...

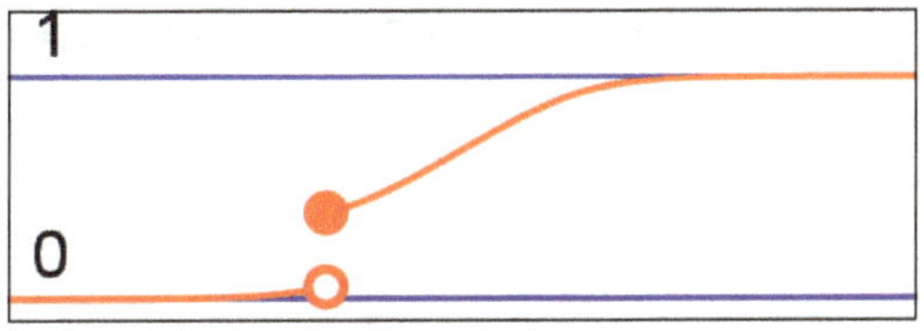

... of a distribution which has both a continuous part and a discrete part.

A discrete probability distribution is a probability distribution characterized by a probability mass function. Thus, the distribution of a random variable X is discrete, and X is called a discrete random variable, if

$$\sum_u \Pr(X = u) = 1$$

as u runs through the set of all possible values of X. A discrete random variable can assume only a finite or countably infinite number of values. For the number of potential values to be countably infinite, even though their probabilities sum to 1, the probabilities have to decline to zero fast enough. For example, if $\Pr(X = n) = \frac{1}{2^n}$ for n = 1, 2, ..., we have the sum of probabilities 1/2 + 1/4 + 1/8 + ... = 1.

Well-known discrete probability distributions used in statistical modeling include the Poisson distribution, the Bernoulli distribution, the binomial distribution, the geometric distribution, and the negative binomial distribution. Additionally, the discrete uniform distribution is commonly used in computer programs that make equal-probability random selections between a number of choices.

Measure Theoretic Formulation

A measurable function $X : A \to B$ between a probability space $(A, \mathcal{A}, P)$ and a measurable space $(B, \mathcal{B})$ is called a discrete random variable provided its image is a countable set and the pre-image of singleton sets are measurable, i.e., $X^{-1}(b) \in \mathcal{A}$ for all $b \in B$. The latter requirement induces a probability mass function $f_X : X(A) \to \mathbb{R}$ via $f_X(b) := P(X^{-1}(b))$. Since the pre-images of disjoint sets are disjoint,

$$\sum_{b \in X(A)} f_X(b) = \sum_{b \in X(A)} P(X^{-1}(b)) = P\left(\bigcup_{b \in X(A)} X^{-1}(b) \right) = P(A) = 1.$$

This recovers the definition given above.

Cumulative Density

Equivalently to the above, a discrete random variable can be defined as a random variable whose cumulative distribution function (cdf) increases only by jump discontinuities—that is, its cdf increases only where it "jumps" to a higher value, and is constant between those jumps. The points where jumps occur are precisely the values which the random variable may take.

Delta-function Representation

Consequently, a discrete probability distribution is often represented as a generalized probability density function involving Dirac delta functions, which substantially unifies the treatment of continuous and discrete distributions. This is especially useful when dealing with probability distributions involving both a continuous and a discrete part.

Indicator-function Representation

For a discrete random variable X, let u_0, u_1, ... be the values it can take with non-zero probability. Denote,

$$\Omega_i = X^{-1}(u_i) = \{\omega : X(\omega) = u_i\}, i = 0,1,2,\ldots$$

These are disjoint sets, and by formula (1),

$$\Pr\left(\bigcup_i \Omega_i\right) = \sum_i \Pr(\Omega_i) = \sum_i \Pr(X = u_i) = 1.$$

It follows that the probability that X takes any value except for u_0, u_1, ... is zero, and thus one can write X as,

$$X(\omega) = \sum_i u_i 1_{\Omega_i}(\omega)$$

except on a set of probability zero, where 1_A is the indicator function of A. This may serve as an alternative definition of discrete random variables.

Continuous Probability Distribution

A continuous probability distribution is a *probability distribution* that has a cumulative distribution function that is continuous. Most often they are generated by having a probability density function. Mathematicians call distributions with probability density functions absolutely continuous, since their cumulative distribution function is absolutely continuous with respect to the Lebesgue measure λ. If the distribution of X is continuous, then X is called a continuous random variable. There are many examples of continuous probability distributions: normal, uniform, chi-squared, and others.

Intuitively, a continuous random variable is the one which can take a continuous range of values—as opposed to a discrete distribution, where the set of possible values for the random variable is at most countable. While for a discrete distribution an event with probability zero is impossible (e.g., rolling $3\frac{1}{2}$ on a standard dice is impossible, and has probability zero), this is not so in the case of a continuous random variable. For example, if one measures the width of an oak leaf, the result of 3½ cm is possible; however, it has probability zero because uncountably many other potential values exist even between 3 cm and 4 cm. Each of these individual outcomes has probability zero, yet the probability that the outcome will fall into the interval (3 cm, 4 cm) is nonzero. This apparent paradox is resolved by the fact that the probability that X attains some value within an infinite set, such as an interval, cannot be found by naively adding the probabilities for individual values. Formally, each value has an infinitesimally small probability, which statistically is equivalent to zero.

Formally, if X is a continuous random variable, then it has a probability density function $f(x)$, and therefore its probability of falling into a given interval, say $[a, b]$ is given by the integral,

$$\Pr[a \le X \le b] = \int_a^b f(x)dx$$

In particular, the probability for X to take any single value a (that is $a \le X \le a$) is zero, because an integral with coinciding upper and lower limits is always equal to zero.

The definition states that a continuous probability distribution must possess a density, or equivalently, its cumulative distribution function be absolutely continuous. This requirement is stronger than simple continuity of the cumulative distribution function, and there is a special class of distributions, singular distributions, which are neither continuous nor discrete nor a mixture of those. An example is given by the Cantor distribution. Such singular distributions however are never encountered in practice.

Note on terminology: some authors use the term "continuous distribution" to denote the distribution with continuous cumulative distribution function. Thus, their definition includes both the (absolutely) continuous and singular distributions.

By one convention, a probability distribution μ is called *continuous* if its cumulative distribution function $F(x) = \mu(-\infty, x]$ is continuous and, therefore, the probability measure of singletons $\mu\{x\}=0$ for all x.

Another convention reserves the term *continuous probability distribution* for absolutely continuous distributions. These distributions can be characterized by a probability density function: a non-negative Lebesgue integrable function f defined on the real numbers such that,

$$F(x) = \mu(-\infty, x] = \int_{-\infty}^{x} f(t)dt.$$

Discrete distributions and some continuous distributions (like the Cantor distribution) do not admit such a density.

Some properties

- The probability distribution of the sum of two independent random variables is the convolution of each of their distributions.
- Probability distributions are not a vector space—they are not closed under linear combinations, as these do not preserve non-negativity or total integral 1—but they are closed under convex combination, thus forming a convex subset of the space of functions (or measures).

Kolmogorov Definition

In the measure-theoretic formalization of probability theory, a random variable is defined as a measurable function X from a probability space $(\Omega, \mathcal{F}, \mathrm{P})$ to measurable space $(\mathcal{X}, \mathcal{A})$. A probability distribution of X is the pushforward measure $X_*\mathrm{P}$ of X, which is a probability measure on $(\mathcal{X}, \mathcal{A})$ satisfying $X_*\mathrm{P} = \mathrm{P}X^{-1}$.

Random Number Generation

A frequent problem in statistical simulations (the Monte Carlo method) is the generation of pseudo-random numbers that are distributed in a given way. Most algorithms are based on a pseudorandom number generator that produces numbers X that are uniformly distributed in the interval [0,1]. These random variates X are then transformed via some algorithm to create a new random variate having the required probability distribution.

Applications

The concept of the probability distribution and the random variables which they describe underlies the mathematical discipline of probability theory, and the science of statistics. There is spread or variability in almost any value that can be measured in a population (e.g. height of people, durability of a metal, sales growth, traffic flow, etc.); almost all measurements are made with some intrinsic error; in physics many processes are described probabilistically, from the kinetic properties of gases to the quantum mechanical description of fundamental particles. For these and many other reasons, simple numbers are often inadequate for describing a quantity, while probability distributions are often more appropriate.

As a more specific example of an application, the cache language models and other statistical language models used in natural language processing to assign probabilities to the occurrence of particular words and word sequences do so by means of probability distributions.

Common Probability Distributions

The following is a list of some of the most common probability distributions, grouped by the type of process that they are related to.

All of the univariate distributions below are singly peaked; that is, it is assumed that the values cluster around a single point. In practice, actually observed quantities may cluster around multiple values. Such quantities can be modeled using a mixture distribution.

Related to Real-valued Quantities that Grow Linearly (e.g. Errors, Offsets)

- Normal distribution (Gaussian distribution), for a single such quantity; the most common continuous distribution

Related to Positive Real-valued Quantities that Grow Exponentially (e.g. Prices, Incomes, Populations)

- Log-normal distribution, for a single such quantity whose log is normally distributed
- Pareto distribution, for a single such quantity whose log is exponentially distributed; the prototypical power law distribution

Related to Real-valued Quantities that are Assumed to be Uniformly Distributed Over a (Possibly Unknown) Region

- Discrete uniform distribution, for a finite set of values (e.g. the outcome of a fair die)
- Continuous uniform distribution, for continuously distributed values

Related to Bernoulli Trials (yes/no Events, with a Given Probability)

- Basic distributions:
  - Bernoulli distribution, for the outcome of a single Bernoulli trial (e.g. success/failure, yes/no)
  - Binomial distribution, for the number of "positive occurrences" (e.g. successes, yes votes, etc.) given a fixed total number of independent occurrences
  - Negative binomial distribution, for binomial-type observations but where the quantity of interest is the number of failures before a given number of successes occurs

  - Geometric distribution, for binomial-type observations but where the quantity of interest is the number of failures before the first success; a special case of the negative binomial distribution

- Related to sampling schemes over a finite population:
  - Hypergeometric distribution, for the number of "positive occurrences" (e.g. successes, yes votes, etc.) given a fixed number of total occurrences, using sampling without replacement
  - Beta-binomial distribution, for the number of "positive occurrences" (e.g. successes, yes votes, etc.) given a fixed number of total occurrences, sampling using a Polya urn scheme (in some sense, the "opposite" of sampling without replacement)

Related to Categorical Outcomes (Events with K Possible Outcomes, with a Given Probability for each Outcome)

- Categorical distribution, for a single categorical outcome (e.g. yes/no/maybe in a survey); a generalization of the Bernoulli distribution
- Multinomial distribution, for the number of each type of categorical outcome, given a fixed number of total outcomes; a generalization of the binomial distribution
- Multivariate hypergeometric distribution, similar to the multinomial distribution, but using sampling without replacement; a generalization of the hypergeometric distribution

Related to Events in a Poisson Process (Events that Occur Independently with a given Rate)

- Poisson distribution, for the number of occurrences of a Poisson-type event in a given period of time
- Exponential distribution, for the time before the next Poisson-type event occurs
- Gamma distribution, for the time before the next k Poisson-type events occur

Related to the Absolute Values of Vectors with Normally Distributed Components

- Rayleigh distribution, for the distribution of vector magnitudes with Gaussian distributed orthogonal components. Rayleigh distributions are found in RF signals with Gaussian real and imaginary components

- Rice distribution, a generalization of the Rayleigh distributions for where there is a stationary background signal component. Found in Rician fading of radio signals due to multipath propagation and in MR images with noise corruption on non-zero NMR signals

Related to Normally Distributed Quantities Operated with Sum of Squares (for Hypothesis Testing)

- Chi-squared distribution, the distribution of a sum of squared standard normal variables; useful e.g. for inference regarding the sample variance of normally distributed samples
- Student's t distribution, the distribution of the ratio of a standard normal variable and the square root of a scaled chi squared variable; useful for inference regarding the mean of normally distributed samples with unknown variance
- F-distribution, the distribution of the ratio of two scaled chi squared variables; useful e.g. for inferences that involve comparing variances or involving R-squared (the squared correlation coefficient)

Useful as Conjugate Prior Distributions in Bayesian Inference

- Beta distribution, for a single probability (real number between 0 and 1); conjugate to the Bernoulli distribution and binomial distribution
- Gamma distribution, for a non-negative scaling parameter; conjugate to the rate parameter of a Poisson distribution or exponential distribution, the precision (inverse variance) of a normal distribution, etc.
- Dirichlet distribution, for a vector of probabilities that must sum to 1; conjugate to the categorical distribution and multinomial distribution; generalization of the beta distribution
- Wishart distribution, for a symmetric non-negative definite matrix; conjugate to the inverse of the covariance matrix of a multivariate normal distribution; generalization of the gamma distribution

Distribution

Probability Distributions

In statistics, a probability distribution gives either the probability of each value of a random variable (when the variable is discrete), or the probability of the value falling within a particular interval (when the variable is continuous). The probability distribution describes the range of possible values that a random variable can attain and the probability that the value of the random variable is within any (measurable) subset of that range.

A probability distribution gives important information about the data, how the values are changing, whether they are bunched together or spread out, and whether they are symmetrically disposed on the X-axis or not. Distribution also tells the relative frequency or proportion of various *X* values in the population in the same way that a histogram gives information about a sample. We now describe the distributions that are commonly used in addressing water resources problems.

Commonly used distributions in hydrology are the Normal, Log Normal, Extreme Value type-1 (Gumbel or EV1), Gamma, Pearson Type - III, and Log Pearson Type - III distributions. A brief description of these distributions is given below.

Normal Distribution

It is also known as the Gaussian distribution. When a hydrologic variable, integrated over a large time period is used in analysis, the variable is expected to follow a normal distribution. The normal distribution has a symmetrical bell-shaped probability density function. The two parameters of the normal distribution are mean μ and standard deviation σ. Its PDF can be expressed as:

$$f(x)=\frac{1}{\sigma\sqrt{2\pi}}\exp\left[-\frac{(x-\mu)^2}{2\sigma^2}\right] \quad -\infty < x < \infty$$

the cumulative density function (CDF) of the normal distribution is:

$$F(x)=\frac{1}{\sigma\sqrt{2\pi}}\int_{-\infty}^{x}\exp\left[-\frac{(x-\mu)^2}{2\sigma^2}\right]du$$

The origin of the normal organization lies in the central limit theorem which states that if a sequence of random variables $x_i, i=1,2......,n$ are independently and identically distributed with mean μ and standard deviation σ then the distribution of n such random variables $Y=\sum_{i=1}^{n} x$ tends to the normal distribution with mean nμ and standard deviation $\sqrt{n}\sigma$ as n becomes large. This theorem holds good irrespective of the probability distribution of x.

The reduced variate of the normal distribution is defined as $Z=(x-\mu)/\sigma$. The properties of the reduced variate are mean = 0, standard deviation $\sigma_z = 1$, and coefficient of skewness = 0. Figure shows the normal distribution and the area under the curve for three values of the reduced variate. As shown, the area under the curve within $\mu \pm \sigma$ is 68.27%, within $\mu \pm 2\sigma$ is 95.45 and it is 99.73 within $\mu \pm 3\sigma$.

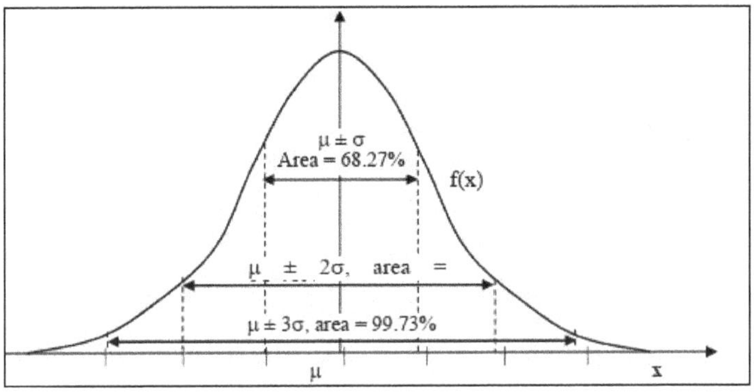

The normal distribution and the area for three values of the standard variate.

Among the probability distributions used in hydrology, the normal distribution is the most widely. It is also employed in the analysis of variance, estimation of random errors of hydrologic measurements, hypothesis testing, synthetic generation of random numbers, etc. A random variable that is made up of the sum of many small independent effects is expected to follow a normal distribution. Many hydrologic variables are not normally distributed, but transformations can, in many cases, make them approximately normally distributed. When there is increase in the time interval over which a hydrologic variable is measured, the variable approximately follows a normal distribution because the number of causative effects increases.

Example: Assuming that the data of Sabarmati River follows the normal distribution, find the parameters of the distribution and plot it.

Solution: For the data of Sabarmati river, the mean and SD are:

Mean of the data $\bar{x} = 665.37$ million cubic m.
Standard deviation $\sigma = 346.9$ million cubic m.
Coefficient of variation $C_V = 346.9/665.37 = 0.521$.
Coefficient of skewness $C_s = 0.76$ (positively skewed).
Kurtosis $C_k = 3.65$.

Figure shows the plot of the probability distribution of Sabarmati River data.

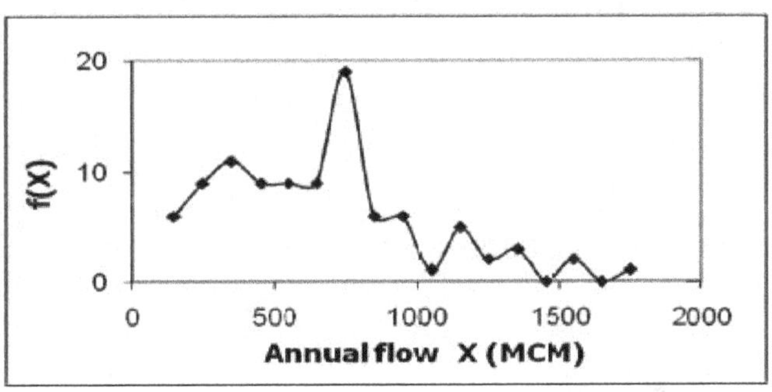

Probability distribution of Sabarmati River data.

For the Sabarmati data, the coefficient of skewness C_s is 0.76 or the data is positively skewed. This is easily verified by figure. Further, kurtosis C_k for the data is 3.65 (kurtosis for the normal distribution is 3). Again, this can also be verified from figure.

Log-Normal Distribution

The log-normal distribution is the probability distribution of a random variable whose logarithm is normally distributed. Let X be a random variable with a normal distribution, then $Y = \exp(X)$ has a log-normal distribution. In other words, if Y is log-normally distributed, then $X = \log(Y)$ is normally distributed. When a random variable represents a process that is the resultant of multiplicative product of many small effects each of which is positive, then it can be expressed the sum of logarithms of these small effects. The logarithm of such a random variable can be expected to follow a normal distribution. Hence, if the variable is transformed to the log domain, it is likely to follow the normal distribution. An advantage of the log-normal distribution is that it is often useful to represent quantities that cannot have negative values. It has proven useful to model rainfall amounts, size distributions of aerosol particles, etc.

The PDF of the log-normal distribution is:

$$f(x) = \frac{1}{x\sigma_y\sqrt{2\pi}} \exp\left[-\frac{(\ln x - \mu_y)^2}{2\sigma_y^2}\right] \quad x > 0$$

The log-normal distribution has two parameters μ_y and σ_y which can be estimated by transforming all x_i's to y_i's by:

$$y_i = \ln x_i$$

Extreme Value Type 1 (EV1) Distribution

Let a series of large number of (N) observations of random variable be subdivided into n subsamples of size m each, such that $N = nm$. Each subseries shall have two extreme values: one maximum and one minimum corresponding to, for example, floods and droughts. Gumbel (1958) showed that the n largest values of subsamples asymptotically follow an extreme value type 1 (EV1) distribution. This distribution, also known as the Gumbel distribution or double negative exponential distribution, is widely used for frequency analysis of floods, maximum rainfall, etc. This distribution is essentially a log-normal distribution with constant skewness (approximately 1.14). Its PDF and CDF are as follows:

$$f(x) = \alpha \exp\{-\alpha(x-\beta) - \exp[-\alpha(x-\beta)]\} \quad -\infty < x < \infty; \quad -\infty < \beta < \infty; \quad \alpha > 0$$

$$F(x) = \exp\{-\exp[-\alpha(x-\beta)]\}$$

where α, and β are scale and location parameters. The estimates of parameters using the method of moments are:

$$\hat{\alpha} = \frac{1.283}{s}; \quad \hat{\beta} = \overline{X} - 0.45s$$

It has been shown that the EVI distribution is a special case of a distribution known as the Generalized Extreme Value (GEV) distribution. The CDF of the GEV distribution is given by:

$$F(x) = \exp\left[-\left(1 - k\frac{x-u}{\alpha}\right)^{1/k}\right]$$

Where k, u, and α are the parameters of the distribution. When k=0, we get the EV1 distribution. For k<0, the distribution known as EV2 and it is known as EV3 distribution when k>0.

According to Gumbel, the probability that an event with magnitude larger than xo occurs is (Subramanya 2008):

$$\text{Prob}\ (X \geq x_o) = 1 - \exp[-\exp(-y)]$$

where y is the reduced variate, given as:

$$y = \alpha(x-\beta),$$
$$\beta = \overline{x} - 0.45\sigma_x$$
$$\alpha = 1.2825 / \sigma_x$$

Substituting the values of a and α

$$y = 1.285(x-\overline{x}) / \sigma_x + 0.577$$

The expression for the reduced variate y for return period T is

$$\text{y} = -[0.834 + 2.303\log(\log\frac{\text{T}}{\text{T}-1})]$$

Now we can compute variate x with return period T by

$$X_T = \overline{x} + k\sigma_x \quad \text{(a)}$$

Where $k = (y_T - 0.577)/1.2825$ (b)

Equations (a) and (b) assume that a large data series are available to compute the various statistics. However in practice, the record length is finite. In such cases, the following equation may be used:

$$X_T = \bar{x} + k_n \sigma_{n-1}$$

where σ_{n-1} is the standard deviation of the sample of size n. Frequency factor for use with sample of size n is given as:

$$k_n = (y_T - \bar{y}_n)/s_n$$

Where $\bar{y}_n$ and s_n are reduced mean and reduced standard deviations which are functions of n. Values of these can be obtained from standard tables that are widely available. Note that as $n \to \infty, \bar{y}_n \to 0.577$ and $s_n \to 1.2825$.

Example: From the flow data of a river, the mean and standard deviation were computed and these turned out to be 660 million cubic m and 330 million cubic m, respectively. Find the value of parameters of EV1 distribution.

Solution: The mean and standard deviation of the data are 665.37 million cubic m and 346.9 million cubic m, respectively. Therefore, the estimates by the method of moment are:

$$\alpha = 1.2825/330 = 0.0039$$

and $\beta = 660 - 0.45 * 330 = 511.5$

Example: Annual maximum flood discharge data of a river was available for 30 years. Mean and standard deviation were 5250 m³/s and 1650 m³/s. Compute the flood discharge with a return period of 100 years by using the Gumbel Extreme Value 1 distribution.

Solution: From standard tables, for n=30 years,

$$y_n = 0.5362, s_n = 1.1124.$$

Hence, $y_T = -[0.834 + 2.303\log(\log\frac{100}{99})] = 4.601$

$$k_{100} = (4.601 - 0.5362)/1.1124 = 3.654$$

$$x_{100} = 5250 + 3.654 * 1650 = 11279 m^3/s$$

Log Pearson Type - III (LP3) Distribution

Log Pearson Type III distribution was found to give good results in numerous studies dealing with flood peak data. This distribution is the standard distribution for flood frequency analysis in the USA since its use for flood frequency analysis was recommended by the US Water Resources Council.

LP3 is a three-parameter distribution and is widely used in hydrology. Its parameters are related to mean, standard deviation, and skewness.

$$f(x)=\frac{1}{a\Gamma(b)}\left(\frac{x-c}{a}\right)^{b-1}\exp\left(-\frac{x-c}{a}\right) \quad \text{(i)}$$

wherea, b, and c are scale, shape, and location parameters, respectively, and $\Gamma(b)$ is a gamma function. If c=0, this distribution becomes a two-parameter gamma distribution. Parameters a, b, and c are related to mean, standard deviation, and coefficient of skewness as (method of moment estimates),

$$a=\sigma/\sqrt{b}$$
$$b=(2/C_s)^2$$
$$c=\mu-\sigma\sqrt{b}$$

To determine flood for a return period T by using the LP3 distribution, the procedure described below is followed.

First of all, the frequency factor, KT is computed by (Chow et al. 1988):

$$K_T=z+(z^2-1)k+(z^3-6z)k^2/3+(z^2-1)k^3+zk^4+k^5/3$$

Where k = $C_s/6$. To complete z for a given return period T, exceedance probability p is obtained as p = 1/T. Now, complete a variable w as:

$$w=\sqrt{\ln(1/P^2)} \qquad 0<P\le 0.5$$

Now z is calculated by (Abramowitz and Stegun, 1965)

$$z=w-\frac{2.515517+0.802853w+0.010328w^2}{1+1.432788w+0.189269w^2+0.001308w^3} \quad \text{(ii)}$$

when p>0.5, p in eq. (ii) is replaced by (1-p) and the negative sign is put before z computed by eq. (iii). Now, by following the frequency factor method, the flood for the return period T years is computed by:

$$y_T=\bar{y}+K_T s_y \quad \text{(iii)}$$

Example: For the data of example, find the parameters of the Pearson Type III distribution.

Solution: The estimates of parameters using the method of moments are,

$$b = (2/C_s)^2 = (2/0.76)^2 = 6.93 \text{ million cubic m.}$$

$$a = 346.9\sqrt{6.93} = 131.78 \text{ million cubic m.}$$

$$c = 665.37 - 346.9 * \sqrt{6.93} = -247.84$$

Example: Logarithms of the annual flood peak data of a river were taken and the mean was 4.146, SD was 0.403 and C_s = -0.07. Find 50 year return period flood by using the LP3 distribution.

Solution First we find the value of K_{50} by the following equation:

$$K_{50} = z + (z^2 - 1)k + (z^3 - 6z)k^2 / 3 + (z^2 - 1)k^3 + zk^4 + k^5 / 3$$

Here $k = C_s / 6 = -0.07 / 6 = -0.0117$. For $T = 50, p = 1/50 = 0.02$.

$$w = \sqrt{\ln(1/0.02^2)} = 2.797$$

From eq. (ii)

$$z = 2.797 - \frac{2.515517 + 0.802853 * 2.797 + 0.010328 * 2.797^2}{1 + 1.432788 * 2.797 + 0.189269 * 2.797^2 + 0.001308 * 2.797^3} = 2.054$$

Now K_{50} is calculated as:

$$K_{50} = 2.054 + \left(2.054^2 - 1\right) * (-0.0117) + \left(2.054^3 - 6 * 2.054\right) * (-0.0117)^2 / 3$$
$$+ \left(2.054^2 - 1\right) * (-0.0117)^3 + 2.054 * (-0.0117)^4 + (-0.0117)^5 / 3$$
$$= 2.016$$

Hence, $y_{50} = 4.146 + 2.016 * 0.403 = 4.959$

So, the 50-year flood $x_{50} = (10)^{4.959} = 90942$.

Discrete Probability Distributions

The use of discrete probability distributions is restricted generally to those random events in which the outcome can be described as success or failure, i.e., there are only two mutually exclusive events in an experiment. Moreover, the successive trials are independent and the probability of success remains constant from trial to trial. The binomial or Poisson distributions can be used to find the probability of occurrence of an event r times in n successive years.

Binomial Distribution

This distribution arises in Bernoulli processes where in any trial; the event may or may not take place. The probability of occurrence of the event is the same from one trial to another. This distribution usually occurs while dealing with complementary events. A common example is tossing of coins in which the probability of head appearing is the same in each trial. The occurrence of wet and dry days over a given time interval is also a complementary event. The probability of occurrence of the event r times in n successive years is given by:

$$P_{r,n} = {}^{n}C_{r} P^{r} q^{n-r} = \frac{n!}{r!(n-r)!} p^{r} q^{n-r}$$

where $P_{r,n}$ is the probability of a random event of a given magnitude and exceedance probability P occurring r times in n successive years. The probability of the event not occurring at all in n successive years is:

$$P_{o,n} = q^{n} = (1-p)^{n}$$

The probability of an event occurring at least once in n successive years:

$$P_{1} = 1 - q^{n} = 1 - (1-p)^{n}$$

Example: An analysis of data on the maximum one-day rainfall depth at a station indicated that a depth of 280 mm had a return period of 50 years. Determine the probability of a one-day rainfall depth equal to or greater than 280 mm occurring (a) once in 20 successive years, and (b) two times in 15 successive years.

Solution: Here, $P = 1/50 = 0.02$.

a) In the first case, $n = 20, r = 1$. Therefore, from equation

$$P_{1,20} = \frac{20!}{19!1!} * (0.02) * (0.98)^{19} = 0.272.$$

b) In this case, $n = 15, r = 2$. Therefore,

$$P_{2,15} = \frac{15!}{13!2!} * (0.02^{2}) * (0.980)^{13} = 0.0292.$$

Example: What is the probability that a 5-year flood will not occur at all in a 10-year period?

Solution: Here, $p = 1/5 = 0.2, n = 10$, and $r = 0$. Hence the probability is

$$P_{0,10} = \frac{10!}{0!10!} * 0.2^0 * (0.8)^{10} = 0.1074$$

Poisson Distribution

The Poisson distribution is a limiting form of the binomial distribution when *p* is very small and n is very large, and *np* tends to a constant value λ. This may happen when the interval over which the Bernoulli process is defined gets smaller and smaller and the number of trials becomes greater and greater, keeping *np* constant. The Poisson distribution has only one parameter λ that denotes the expected mean frequency of occurrence of some event in a given time t. The probability distribution of the number of events in a given time is,

$$P(X = x) = \frac{\lambda^x \exp(-\lambda)}{x!}, \lambda > 0, \ x = 0,1,2...$$

The CDF of the Poisson distribution is,

$$P(X \leq x) = \sum_{i=0}^{x} \frac{\lambda^i \exp(-\lambda)}{i!}$$

The conditions for application of Poisson distribution are: a) the number of events is discrete, b) two events cannot coincide, c) the mean number of events per unit time is constant, and d) events are independent. Thus, it can be applied to following situations with p relatively small and n relatively large to determine the probability of:

(i) Droughts in a given time period,

(ii) Number of rainy days at a given location,

(iii) Probability of rare flood events,

(iv) Probability of reservoir being empty in any one year out of a long period of record.

Parameter Estimation Methods

A number of methods have been developed to estimate parameters of hydrologic models. Some commonly used methods in hydrology include: (1) method of moments; (2) method of probability weighted moments; (4) L-moments; (5) maximum likelihood estimation; and (6) least squares method. Each of these methods is discussed here.

Method of Moments

This method is very commonly employed to estimate parameters of linear hydrologic models. This method is based on the premise that when the parameters of a probability distribution are estimated correctly, the moments of the probability density function are equal to the corresponding moments of a sample data. Nash (1959) developed the theorem of moments which relates the moments of input, output and impulse response functions of linear hydrologic models.

Let X be a continuous variable and $f(x)$ its function satisfying some necessary conditions. The rth moment of $f(x)$ about an arbitrary point 'a' is denoted as $M_r^a(f)$. The rth moment of the function $f(x)$ can be expressed as:

$$M_r^a(f) = \int_{-\infty}^{\infty} (x-a)^r f(x)dx$$

Figure shows the definition of various terms used in the above equation.

Consider the special case when r = 0. In this case, the above eq. gives

$$M_0^a = \int_{-\infty}^{\infty} (x-a)^0 f(x)dx = \int_{-\infty}^{\infty} f(x)dx = 1$$

Thus, the zero-order moment is the area under the curve defined by $f(x)$ subject to $-\infty < x < \infty$. For probability distribution, this area is unity. If $r = 1$, then

$$M_1^a = \int_{-\infty}^{\infty} (x-a)^1 f(x)dx = \mu - a$$

where μ is the mean. If the moment is taken around the origin, then $a = 0$, and the first moment gives the mean. When $a = \mu$, the rth moment about the mean is expressed by:

$$M_r^{\mu} = \int_{-\infty}^{\infty} (x-\mu)^r f(x)dx$$

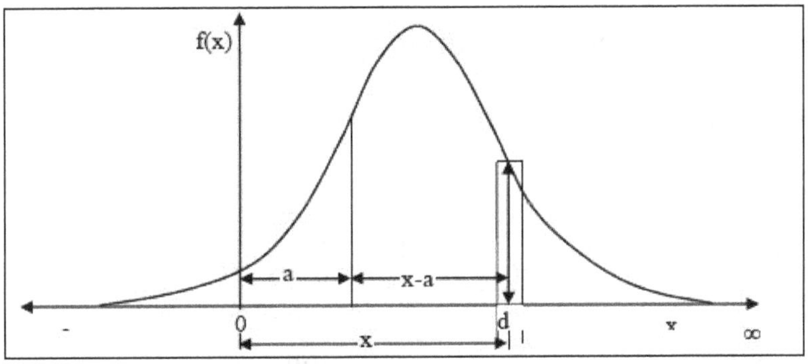

Concept of moment of a function f(x) about an arbitrary point.

For simplicity of notation, we drop the superscript if the moment is taken about the origin o and the familiar terminology of the moments can be written as follows:

M_0 =Area,

M_1 = Mean,

M_2^{μ} = Variance,

M_3^{μ} = Measurement of skewness of the function,

M_4^{μ} = Kurtosis,

Method of Moments for Discrete Systems

For a discrete function, represented as $f_{j,} j = -\infty, ..., -1, 0, 1, ..., \infty$, the rth moment about any arbitrary point can be defined in an analogous manner as for continuous functions. The rth moment about the origin, is defined as:

$$M_r = \sum_{m=-\infty}^{\infty} m^r f_m$$

Figure depicts the concept of moment of a discrete function.

Example: The frequency table of annual flows of Sabarmati River is given table. Find the mean and variance of the data by using the method of moments.

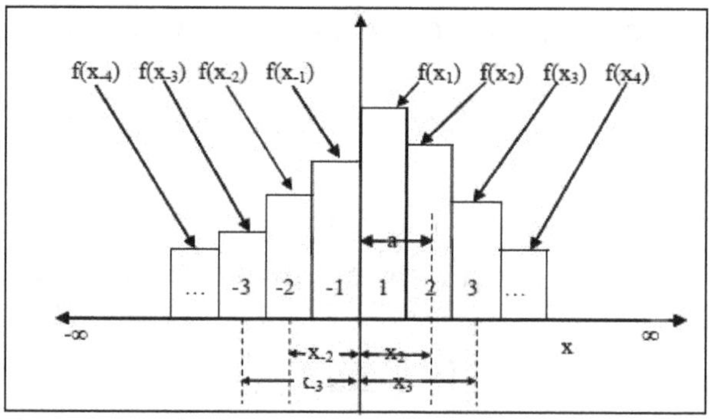

Concept of moment of a discrete function about an arbitrary point 'a'.

Table: Frequency table of annual flows of Sabarmati River.

Discharge range	Frequency	Discharge range	Frequency
100-200	6	200-300	9
300-400	11	400-500	9
500-600	9	600-700	9
700-800	19	800-900	6
900-1000	6	1000-1100	1
1100-1200	5	1200-1300	2
1300-1400	3	1400-1500	0
1500-1600	2	1600-1700	0
1700-1800	1		

Solution: The sum of all frequencies = (6+11+9+19+ ...) = 98. For the first range of discharge (100-200), the mean value is 150; for the second range (200-300), it is 250 and so on.

Hence, the first moment of the data = (150*6 + 250*9 + 350*11 + ...+ 1750*1)/98

= 664.29 cumec.

This is the mean (x_m) of the annual flows.

The second moment about the mean gives the variance of the data.

Second $M_2^{\mu} = \sum_{i=1}^{N} \frac{(x_i - x_m)^2 f(x_i)}{N}$ *moment*

$= [(150\text{-}664)^2 * 6 + (250 - 664)^2 * 9 + \ldots + (1750\text{-}664)^2 * 1]/98$

= 120000 cumec^2.

Hence, the standard deviation (s) = $(120000)^{0.5}$ =346.61 cumec.

This leads to the coefficient of variation $c_v = s/x_m$ =346.61/664.29 = 0.52.

Method of Maximum Likelihood (MLE)

The maximum likelihood (ML) estimation method is widely accepted as one of the most powerful parameter estimation methods. Asymptotically, the ML parameter estimates are unbiased, have minimum variance, and are normally distributed, while in some cases these properties hold for small samples. The MLE method has been extensively used for estimating parameters of frequency distributions as well as fitting conceptual models.

Let $f(x; a_1, a_2, \ldots a_m)$ be a PDF of the random variable X with parameters a_i, i=1, 2, ..., m, to be estimated. For a random sample of data $x_1, x_2, \ldots x_n$, drawn from this probability density function, the joint PDF is defined as:

$$f(x_i, x_2, \ldots . x_n; a_1, a_2, \ldots . a_m) = \underset{i=1}{\overset{n}{\pi}} f(x_i; a_1, a_2, \ldots .. a_m)$$

Interpreted conceptually, the probability of obtaining a given value of X, say x_1, is proportional to $f(x; a_1, a_2, \ldots a_m)$. Likewise, the probability of obtaining the random sample $x_1, x_2, \ldots x_n$ from the population of X is proportional to the product of the individual probability densities or the joint PDF. This joint PDF is called the likelihood function, denoted by L.

$$L = \underset{i=1}{\overset{n}{\pi}} f(x_1; a_1, a_2, \ldots .. a_m)$$

where the parameters a_i, i=1,2,...m, are unknown.

By maximizing the likelihood that the sample under consideration is the one that would be obtained if n random observations were selected from $f(x; a_1, a_2, \dots a_m)$, the unknown parameters are determined, and hence the name of the method. The values of parameters so obtained are known as MLE estimators. Since the logarithm of L attains its maximum for the same values of ai, $i = 1, 2, \dots m$, as does L, the MLE function can also be expressed as:

$$\ln L = L^* = \ln \underset{i=1}{\overset{n}{\pi}} f(x_1; a_1, a_2, \dots. a_m) = \sum_{i=1}^{n} \ln f(x_1; a_1, a_2, \dots. a_m)$$

Frequently ln[L] is maximized, for it is many times easier to find the maximum of the logarithm of the maximum likelihood function than that of the normal L.

The procedure for estimating parameters or determining the point where the MLE function achieves its maximum involves differentiating L or $ln\ L$ partially with respect to each parameter and equating each differential to zero. This results in as many equations as the number of unknown parameters. For m unknown parameters, we get,

$$\frac{\partial L(a_1, a_2, \dots a_m)}{\partial a_1} = 0$$

$$\frac{\partial L(a_1, a_2, \dots a_m)}{\partial a_2} = 0$$

$$\frac{\partial L(a_1, a_2, \dots a_m)}{\partial a_m} = 0$$

These m equations in m unknowns are then solved for the m unknown parameters.

Example: Using the method of maximum likelihood, find the parameter α of the exponential distribution for the data of the Sabarmati River in India, given in example.

Solution: The probability density function of the one-parameter exponential distribution is given by,

$$fx(x) = \alpha exp(-\alpha x)$$

The likelihood function is given by,

$$L(\alpha) = \prod_{i=1}^{n} \alpha \exp(-\alpha x_i) = \alpha^n \exp\left(-\alpha \sum_{i=1}^{n} x_i\right)$$

This can be used to form the log-likelihood function:

$$\ln L(\alpha) = n \ln(\alpha) - \alpha\left(\sum_{i=1}^{n} x_i\right)$$

where n is the sample size. Differentiating above equation with respect to α:

$$\frac{d\ln L(\alpha)}{d\alpha} = \frac{n}{\alpha} - \sum_{i=1}^{n} x_i = 0$$

This yields,

$$\alpha = n \Big/ \left(\sum_{i=1}^{n} x_i \right) = \frac{1}{\bar{x}}$$

In example, the mean of the data was found to be 664.29 cumec. Hence, the estimate of α is:

$$\alpha = 1/664.29 = 1.51 \times 10^{-3} \text{ cumec}^{-1}.$$

Method of Least Squares

The method of least squares (MOLS) is one of the most frequently used parameter estimation methods in hydrology. Natale and Todini (1974) presented constrained MOLS for linear models in hydrology.

Let there be a function $Y = f(X; a_1, a_2, \ldots a_m)$, where a_i, $i = 1, 2, \ldots m$, are parameters to be estimated. The method of least squares (MOLS) involves estimating parameters by minimizing the sum of squares of all deviations between observed and computed values of *Y*. Mathematically, this sum *D* can be expressed as:

$$D = \sum_{i=1}^{n} d_1^2 = \sum_{i=1}^{n} [y_0(i) - y_c(i)]^2 = \sum_{i=1}^{n} [y_0(i) - f(x; a_1, a_2, \ldots a_m)]^2$$

where $y_0(i)$ is the ith observed value of Y, $y_c(i)$ is the ith computed value of Y, and $n > m$ is the number of observations. The minimum of D in equation can be obtained by differentiating D partially with respect to each parameter and equating each differential to zero, e.g.,

$$\frac{\partial \sum_{i=1}^{n} [y_0(i) - f(x_i; a_1, a_2, \ldots a_m)]^2}{\partial a_1} = 0$$

The resulting m equations, usually called the normal equations, are then solved for estimation of m parameters. This method is frequently used to estimate parameters of linear regression model.

Method of L-Moments

Greenwood et al. (1979) introduced the method of probability weighted moments (PWM) and showed its usefulness in deriving explicit expressions for parameters of distributions whose inverse forms X=X (F) can be explicitly defined. They derived relations between parameters and PWMs for Generalized Lambda, Wakeby, Weibull, Gumbel, Logistic and Kappa distributions. However, the probability-weighted moments characterize a distribution but are not meaningful by themselves.

L-moments were developed by Hosking (1986) as functions of PWMs which provide a descriptive summary of the location, scale, and shape of the probability distribution. These moments are analogous to ordinary moments and are expressed as *linear* combinations of order statistics, hence the name. They can also be expressed by linear combinations of probability- weighted moments. Thus, the ordinary moments, the probability weighted moments, and L- moments are related to each other. L-moments are known to have several important advantages over ordinary moments. L-moments have less bias than ordinary moments because they are linear combinations of ranked observations. As an example, variance (second moment) and skewness (third moment) involve squaring and cubing of observations, respectively, which compel them to give greater weight to the observations far from the mean. As a result, they result in substantial bias and variance.

The first L-moment denoted as λ_1 is the arithmetic mean:

$$\lambda_1 = E[X]$$

Let us consider a sample of size n and arrange the data such that $X_{(i|n)}$ is the ith largest observation; clearly i = n will be the largest value. Then, for any distribution, the second L- moment, λ_2, is a description of scale based upon the expected difference between two randomly selected observations:

$$\lambda_2 = (1/2)\ E[X_{(2|1)} - X_{(1|2)}]$$

To compute L-moment measures of skewness three randomly selected observations are used and for kurtosis, we use four randomly selected observations.

$$\lambda_3 = (1/3)\ E[X_{(3|3)} - 2X_{(2|3)} + X_{(1|3)}]$$

$$\lambda_4 = (1/4)\ E[X_{(4|4)} - 3X_{(3|4)} + 3X_{(2|4)} - X_{(1|4)}]$$

Sample L-moment estimates are often computed using (PWMs). The rth PWM is defined (Loucks and Beek, 2005) as:

$$\beta_r = E\{X\,[F(X)]^r\}$$

where $F(X)$ is the cumulative distribution function of X. Recommended (Landwehr et al., 1979; Hosking and Wallis, 1995) unbiased PWM estimators, b_r, of β_r are computed as:

$$b_0 = \overline{X}$$

$$b_1 = \frac{1}{n(n-1)}\sum_{j=2}^{n}(j-1)X_{(j)}$$

$$b_2 = \frac{1}{n(n-1)(n-2)}\sum_{j=3}^{n}(j-1)(j-2)X_{(j)}$$

$$b_3 = \frac{1}{n(n-1)(n-2)(n-3)}\sum_{j=4}^{n}(j-1)(j-2)(j-3)X_{(j)}$$

The general formula for computing estimators b_r of β_r is given by,

$$b_r = \frac{1}{n}\sum_{j=r+1}^{n}\binom{j-1}{r}X_{(j)} \Big/ \binom{n-1}{r}$$

$$= \frac{1}{r+1}\sum_{j=r+1}^{n}\binom{j-1}{r}X_{(j)} \Big/ \binom{n}{r+1}$$

for $r = 1, \ldots, n\text{-}1$.

L-moments are easily calculated in terms of PWMs using:

$$\lambda_1 = \beta_0$$

$$\lambda_2 = 2\beta_1 - \beta_0$$

$$\lambda_3 = 6\beta_2 - 6\beta_1 + \beta_0$$

$$\lambda_4 = 20\beta_3 - 30\beta_2 + 12\beta_1 - \beta_0$$

As with traditional product moments, measures of the coefficient of variation, skewness and kurtosis of a distribution can be computed with L-moments. Following L-moment ratios are important:

L- coefficient of variation (L-CV)	$t_2 = \lambda_2 / \lambda_1$
L- coefficient of skewness (L-sk)	$t_3 = \lambda_3 / \lambda_2$
L- coefficient of kurtosis (L-ku)	$t_4 = \lambda_4 / \lambda_2$

Example: Table gives annual discharge data of a river for 36 years. Compute sample L-moments and L-moment ratios, L-CV, L-sk, and L-ku.

Table: Annual discharge data of a river for 36 years

Year	Discharge	Year	Discharge	Year	Discharge	Year	Discharge
1950	400	1959	1390	1968	2291	1977	1499
1951	1100	1960	3300	1969	1340	1978	2598
1952	900	1961	2190	1970	3200	1979	3487
1953	440	1962	935	1971	2200	1980	1234
1954	3000	1963	785	1972	1014	1981	819
1955	2500	1964	501	1973	1790	1982	1210
1956	760	1965	1123	1974	1140	1983	1510
1957	1250	1966	1581	1975	764	1984	1780
1958	1340	1967	959	1976	783	1985	1398

Solution: Yields estimates of the first three Probability Weighted Moments:

$$b_0 = 1514.19$$
$$b_1 = 889.16$$
$$b_2 = 655.38$$
$$b_3 = 518.64$$

The sample L-moments can be calculated using the probability weighted moments to obtain:

$$\hat{\lambda}_1 = b_0 = 1514.19$$

$$\hat{\lambda}_2 = 2b_1 - b_0 = 264.12$$

$$\hat{\lambda}_3 = 6b_2 - 6b_1 + b_0 = 111.53$$

$$\hat{\lambda}_4 = 20b_3 - 30b_2 + 12b_1 - b_0 = -132.82$$

Thus, the sample estimates of the L-coefficient of variation, t_2, and L-coefficient of skewness, t_3, are:

$$t_2 = 264.12/1514.19 = 0.174$$
$$t_3 = 111.53/264.12 = 0.422$$
$$t_4 = -132.82/264.12 = -0.502$$

Problems of Parameter Estimation

The parameters of a distribution function are estimated from the available sample data. But while doing so, errors may arise due to many reasons. The sample data may contain errors, the assumption underlying a particular method of parameter estimation may not hold good, and there may be truncation and round-off errors. All these may result in errors in estimates of parameter. Each estimate of a parameter is a function of sample parameter data which are observations of a random variable. Thus, the estimate value of the parameter itself is a random variable with certain distribution. An estimate obtained from a given set of values can be regarded as an observed value of the random variable. Thus, the goodness of an estimate can be judged from its distribution.

Several questions arise in parameter estimation. How should we employ the available data to obtain the best estimate? What is the best estimate? Are these estimates unique? A number of statistical properties are available by which to address the above questions. These are discussed below.

Bias

Let the estimate of parameter *a* be a_c denoted by. Estimate a_c will be called an unbiased estimate of *a* if the expected value of a, denoted $E(a_c) = a$. In general, an estimate will have a certain bias *b(a)* depending on *a* so that,

$$E(a_c) = a + b$$

An unbiased estimate mean $b(a) = 0$. Note that an individual a_c may not be equal to or close to *a* even if $b(a) = 0$. Unbiasedness simply implies that the average of many independent estimates of *a* will be equal to *a*.

The bias in a given quantity is usually measured in dimensionless terms and is often referred to as standardized bias (or BIAS). Thus, BIAS is defined as:

$$BIAS = \frac{E(\hat{a}) - a}{a}$$

where â is an estimate of parameter or quantile of *a*. In Monte Carlo experimentation, large numbers of samples of different sizes are generated from a given population. For each sample, then, an estimate of *a* is obtained. If there are, say, 1000 samples of a given size generated then there are 1000 values of parameter *a*. Thus, earlier equation is the average of the 1000 estimates of *a* for a given sample size and is estimated as:

$$E(\hat{a}) = \sum_{i=1}^{n} \hat{a}_i / n$$

where n is the number of samples generated or the number of values of the a estimate. The value of a in eq. $BIAS = \frac{E(\hat{a})-a}{a}$ is the true value of a or the value of parameter a of the population.

Efficiency

An estimate ac of a is said to be efficient if it is unbiased and its variance is at least as small as that of any other unbiased estimate of a. If there are two estimates of a, say a_1 and a_2, then the relative efficiency of a_1 with respect to a_2 is defined as:

$$e = \frac{E(a_1 - a)^2}{E(a_2 - a)^2} \leq 1$$

if $E(a_2\text{-}a)^2 > E(a_1\text{-}a)^2$, then $e \leq 1$. An efficient estimate has e = 1. If an efficient estimate exists, it may be approximately obtained by use of the MLE or entropy method.

Standard Error

Another dimensionless performance measure frequently used in hydrology is the standard error (SE), defined as:

$$SE = \sigma(\hat{a})/a$$

where σ (.) denotes the standard deviation of a and is computed as:

$$\sigma(\hat{a}) = \left[\frac{1}{n-1}\sum_{i=1}^{n}\{\hat{a}_i - E(\hat{a}_i)\}^2\right]^{1/2}$$

where the summations are over n estimates $\hat{a}$ of a. In Monte Carlo experiments, referred to as above, for each sample size, a value of SE is obtained. Thus, this measure is similar to the coefficient of variation.

Root Mean Square Error

The root mean square error (RMSE) is one of the most frequently employed performance measures and is defined for parameter a estimate as:

$$RMSE = E\left[(\hat{a} - a)^2\right]^{1/2} / a$$

where E[.] is the expectation of [.]. It can be shown that RMSE is related to BIAS and SE as:

$$RMSE = \left[\frac{n-1}{n}SE^2 + BIAS^2\right]^{1/2}$$

Relative Mean Error

Another measure of error in assessing the goodness of fit of hydrologic models is the relative mean error (RME) defined as:

$$RME = \frac{1}{N}\left(\sum_{i=1}^{N}\left[\frac{Q_0 - Q_c}{Q_0}\right]^2\right)^{0.5}$$

in which N is the sample size, Q is the observed quantity of a given probability and Q_c is the computed quantity of the same probability. Also, used sometimes is the relative absolute error defined as:

$$RAE = \frac{1}{N}\sum_{i=1}^{N}\left|\frac{Q_0 - Q_c}{Q_c}\right|$$

Statistical Hypothesis Testing

A statistical hypothesis, sometimes called confirmatory data analysis, is a hypothesis that is testable on the basis of observing a process that is modeled via a set of random variables. A statistical hypothesis test is a method of statistical inference. Commonly, two statistical data sets are compared, or a data set obtained by sampling is compared against a synthetic data set from an idealized model. A hypothesis is proposed for the statistical relationship between the two data sets, and this is compared as an alternative to an idealized null hypothesis that proposes no relationship between two data sets. The comparison is deemed *statistically significant* if the relationship between the data sets would be an unlikely realization of the null hypothesis according to a threshold probability—the significance level. Hypothesis tests are used in determining what outcomes of a study would lead to a rejection of the null hypothesis for a pre-specified level of significance. The process of distinguishing between the null hypothesis and the alternative hypothesis is aided by identifying two conceptual types of errors (type 1 & type 2), and by specifying parametric limits on e.g. how much type 1 error will be permitted.

An alternative framework for statistical hypothesis testing is to specify a set of statistical models, one for each candidate hypothesis, and then use model selection techniques to choose the most appropriate model. The most common selection techniques are based on either Akaike information criterion or Bayes factor.

Confirmatory data analysis can be contrasted with exploratory data analysis, which may not have pre-specified hypotheses.

Variations and Sub-classes

Statistical hypothesis testing is a key technique of both frequentist inference and Bayesian inference, although the two types of inference have notable differences. Statistical hypothesis tests define a procedure that controls (fixes) the probability of incorrectly *deciding* that a default position (null hypothesis) is incorrect. The procedure is based on how likely it would be for a set of observations to occur if the null hypothesis were true. Note that this probability of making an incorrect decision is *not* the probability that the null hypothesis is true, nor whether any specific alternative hypothesis is true. This contrasts with other possible techniques of decision theory in which the null and alternative hypothesis are treated on a more equal basis.

One naïve Bayesian approach to hypothesis testing is to base decisions on the posterior probability, but this fails when comparing point and continuous hypotheses. Other approaches to decision making, such as Bayesian decision theory, attempt to balance the consequences of incorrect decisions across all possibilities, rather than concentrating on a single null hypothesis. A number of other approaches to reaching a decision based on data are available via decision theory and optimal decisions, some of which have desirable properties. Hypothesis testing, though, is a dominant approach to data analysis in many fields of science. Extensions to the theory of hypothesis testing include the study of the power of tests, i.e. the probability of correctly rejecting the null hypothesis given that it is false. Such considerations can be used for the purpose of sample size determination prior to the collection of data.

The Testing Process

In the statistics literature, statistical hypothesis testing plays a fundamental role. The usual line of reasoning is as follows:

1. There is an initial research hypothesis of which the truth is unknown.
2. The first step is to state the relevant null and alternative hypotheses. This is important, as mis-stating the hypotheses will muddy the rest of the process.
3. The second step is to consider the statistical assumptions being made about the sample in doing the test; for example, assumptions about the statistical independence or about the form of the distributions of the observations. This is equally important as invalid assumptions will mean that the results of the test are invalid.
4. Decide which test is appropriate, and state the relevant test statistic T.
5. Derive the distribution of the test statistic under the null hypothesis from the assumptions. In standard cases this will be a well-known result. For example, the test statistic might follow a Student's t distribution or a normal distribution.

6. Select a significance level (α), a probability threshold below which the null hypothesis will be rejected. Common values are 5% and 1%.
7. The distribution of the test statistic under the null hypothesis partitions the possible values of T into those for which the null hypothesis is rejected—the so-called *critical region*—and those for which it is not. The probability of the critical region is α.
8. Compute from the observations the observed value t_{obs} of the test statistic T.
9. Decide to either reject the null hypothesis in favor of the alternative or not reject it. The decision rule is to reject the null hypothesis H_0 if the observed value t_{obs} is in the critical region, and to accept or "fail to reject" the hypothesis otherwise.

An alternative process is commonly used:

1. Compute from the observations the observed value t_{obs} of the test statistic T.
2. Calculate the p-value. This is the probability, under the null hypothesis, of sampling a test statistic at least as extreme as that which was observed.
3. Reject the null hypothesis, in favor of the alternative hypothesis, if and only if the p-value is less than the significance level (the selected probability) threshold.

The two processes are equivalent. The former process was advantageous in the past when only tables of test statistics at common probability thresholds were available. It allowed a decision to be made without the calculation of a probability. It was adequate for classwork and for operational use, but it was deficient for reporting results.

The latter process relied on extensive tables or on computational support not always available. The explicit calculation of a probability is useful for reporting. The calculations are now trivially performed with appropriate software.

The difference in the two processes applied to the Radioactive suitcase example (below):

- "The Geiger-counter reading is 10. The limit is 9. Check the suitcase."
- "The Geiger-counter reading is high; 97% of safe suitcases have lower readings. The limit is 95%. Check the suitcase."

The former report is adequate, the latter gives a more detailed explanation of the data and the reason why the suitcase is being checked.

It is important to note the difference between accepting the null hypothesis and simply failing to reject it. The "fail to reject" terminology highlights the fact that the null hypothesis is assumed to be true from the start of the test; if there is a lack of evidence against it, it simply continues to be assumed true. The phrase "accept the null hypothesis" may suggest it has been proved simply because it has not been disproved, a logical fallacy known as the argument from ignorance. Unless a test with particularly high power is used, the idea of "accepting" the null hypothesis may be dangerous. Nonetheless the terminology is prevalent throughout statistics, where the meaning actually intended is well understood.

The processes described here are perfectly adequate for computation. They seriously neglect the design of experiments considerations.

It is particularly critical that appropriate sample sizes be estimated before conducting the experiment.

The phrase "test of significance" was coined by statistician Ronald Fisher.

Interpretation

If the *p*-value is less than the required significance level (equivalently, if the observed test statistic is in the critical region), then we say the null hypothesis is rejected at the given level of significance. Rejection of the null hypothesis is a conclusion. This is like a "guilty" verdict in a criminal trial: the evidence is sufficient to reject innocence, thus proving guilt. We might accept the alternative hypothesis (and the research hypothesis).

If the *p*-value is *not* less than the required significance level (equivalently, if the observed test statistic is outside the critical region), then the test has no result. The evidence is insufficient to support a conclusion. (This is like a jury that fails to reach a verdict.) The researcher typically gives extra consideration to those cases where the *p*-value is close to the significance level.

Some people find it helpful to think of the hypothesis testing framework as analogous to a mathematical proof by contradiction.

In the Lady tasting tea example, Fisher required the Lady to properly categorize all of the cups of tea to justify the conclusion that the result was unlikely to result from chance. He defined the critical region as that case alone. The region was defined by a probability (that the null hypothesis was correct) of less than 5%.

Whether rejection of the null hypothesis truly justifies acceptance of the research hypothesis depends on the structure of the hypotheses. Rejecting the hypothesis that a large paw print originated from a bear does not immediately prove the existence of Bigfoot. Hypothesis testing emphasizes the rejection, which is based on a probability, rather than the acceptance, which requires extra steps of logic.

"The probability of rejecting the null hypothesis is a function of five factors: whether the test is one- or two tailed, the level of significance, the standard deviation, the amount of deviation from the null hypothesis, and the number of observations." These factors are a source of criticism; factors under the control of the experimenter/analyst give the results an appearance of subjectivity.

Use and Importance

Statistics are helpful in analyzing most collections of data. This is equally true of hypothesis testing which can justify conclusions even when no scientific theory exists. In the Lady tasting tea example, it was "obvious" that no difference existed between (milk poured into tea) and (tea poured into milk). The data contradicted the "obvious".

Real world applications of hypothesis testing include:

- Testing whether more men than women suffer from nightmares
- Establishing authorship of documents
- Evaluating the effect of the full moon on behavior
- Determining the range at which a bat can detect an insect by echo
- Deciding whether hospital carpeting results in more infections
- Selecting the best means to stop smoking
- Checking whether bumper stickers reflect car owner behavior
- Testing the claims of handwriting analysts

Statistical hypothesis testing plays an important role in the whole of statistics and in statistical inference. For example, Lehmann (1992) in a review of the fundamental paper by Neyman and Pearson (1933) says: "Nevertheless, despite their shortcomings, the new paradigm formulated in the 1933 paper, and the many developments carried out within its framework continue to play a central role in both the theory and practice of statistics and can be expected to do so in the foreseeable future".

Significance testing has been the favored statistical tool in some experimental social sciences (over 90% of articles in the *Journal of Applied Psychology* during the early 1990s). Other fields have favored the estimation of parameters (e.g., effect size). Significance testing is used as a substitute for the traditional comparison of predicted value and experimental result at the core of the scientific method. When theory is only capable of predicting the sign of a relationship, a directional (one-sided) hypothesis test can be configured so that only a statistically significant result supports theory. This form of theory appraisal is the most heavily criticized application of hypothesis testing.

Cautions

"If the government required statistical procedures to carry warning labels like those on drugs, most inference methods would have long labels indeed." This caution applies to hypothesis tests and alternatives to them.

The successful hypothesis test is associated with a probability and a type-I error rate. The conclusion *might* be wrong.

The conclusion of the test is only as solid as the sample upon which it is based. The design of the experiment is critical. A number of unexpected effects have been observed including:

- The clever Hans effect. A horse appeared to be capable of doing simple arithmetic.
- The Hawthorne effect. Industrial workers were more productive in better illumination, and most productive in worse.
- The placebo effect. Pills with no medically active ingredients were remarkably effective.

A statistical analysis of misleading data produces misleading conclusions. The issue of data quality can be more subtle. In forecasting for example, there is no agreement on a measure of forecast accuracy. In the absence of a consensus measurement, no decision based on measurements will be without controversy.

The book *How to Lie with Statistics* is the most popular book on statistics ever published. It does not much consider hypothesis testing, but its cautions are applicable, including: Many claims are made on the basis of samples too small to convince. If a report does not mention sample size, be doubtful.

Hypothesis testing acts as a filter of statistical conclusions; only those results meeting a probability threshold are publishable. Economics also acts as a publication filter; only those results favorable to the author and funding source may be submitted for publication. The impact of filtering on publication is termed publication bias. A related problem is that of multiple testing (sometimes linked to data mining), in which a variety of tests for a variety of possible effects are applied to a single data set and only those yielding a significant result are reported. These are often dealt with by using multiplicity correction procedures that control the family wise error rate (FWER) or the false discovery rate (FDR).

Those making critical decisions based on the results of a hypothesis test are prudent to look at the details rather than the conclusion alone. In the physical sciences most results are fully accepted only when independently confirmed. The general advice concerning statistics is, "Figures never lie, but liars figure".

Examples

Lady Tasting Tea

In a famous example of hypothesis testing, known as the *Lady tasting tea*, Dr. Muriel Bristol, a female colleague of Fisher claimed to be able to tell whether the tea or the milk was added first to a cup. Fisher proposed to give her eight cups, four of each variety, in random order. One could then ask what the probability was for her getting the number she got correct, but just by chance. The null hypothesis was that the Lady had no such ability. The test statistic was a simple count of the number of successes in selecting the 4 cups. The critical region was the single case of 4 successes of 4 possible based on a conventional probability criterion (< 5%; 1 of 70 ≈ 1.4%). Fisher asserted that no alternative hypothesis was (ever) required. The lady correctly identified every cup, which would be considered a statistically significant result.

Courtroom Trial

A statistical test procedure is comparable to a criminal trial; a defendant is considered not guilty as long as his or her guilt is not proven. The prosecutor tries to prove the guilt of the defendant. Only when there is enough charging evidence the defendant is convicted.

In the start of the procedure, there are two hypotheses H_0: "the defendant is not guilty", and H_1: "the defendant is guilty". The first one, H_0, is called the *null hypothesis*, and is for the time being accepted. The second one, H_1, is called the *alternative hypothesis*. It is the alternative hypothesis that one hopes to support.

The hypothesis of innocence is only rejected when an error is very unlikely, because one doesn't want to convict an innocent defendant. Such an error is called *error of the first kind* (i.e., the conviction of an innocent person), and the occurrence of this error is controlled to be rare. As a consequence of this asymmetric behaviour, the *error of the second kind* (acquitting a person who committed the crime), is often rather large.

	H_0 is true Truly not guilty	H_1 is true Truly guilty
Accept null hypothesis Acquittal	Right decision	Wrong decision Type II Error
Reject null hypothesis Conviction	Wrong decision Type I Error	Right decision

A criminal trial can be regarded as either or both of two decision processes: guilty vs not guilty or evidence vs a threshold ("beyond a reasonable doubt"). In one view, the defendant is judged; in the other view the performance of the prosecution (which bears the burden of proof) is judged. A hypothesis test can be regarded as either a judgment of a hypothesis or as a judgment of evidence.

Philosopher's Beans

The following example was produced by a philosopher describing scientific methods generations before hypothesis testing was formalized and popularized.

Few beans of this handful are white. Most beans in this bag are white. Therefore: Probably, these beans were taken from another bag. This is an hypothetical inference.

The beans in the bag are the population. The handful are the sample. The null hypothesis is that the sample originated from the population. The criterion for rejecting the null-hypothesis is the "obvious" difference in appearance (an informal difference in the mean). The interesting result is that consideration of a real population and a real sample produced an imaginary bag. The philosopher was considering logic rather than probability. To be a real statistical hypothesis test, this example requires the formalities of a probability calculation and a comparison of that probability to a standard.

A simple generalization of the example considers a mixed bag of beans and a handful that contain either very few or very many white beans. The generalization considers both extremes. It requires more calculations and more comparisons to arrive at a formal answer, but the core philosophy is unchanged; If the composition of the handful is greatly different from that of the bag, then the sample probably originated from another bag. The original example is termed a one-sided or a one-tailed test while the generalization is termed a two-sided or two-tailed test.

The statement also relies on the inference that the sampling was random. If someone had been picking through the bag to find white beans, then it would explain why the handful had so many white beans, and also explain why the number of white beans in the bag was depleted (although the bag is probably intended to be assumed much larger than one's hand).

Clairvoyant Card Game

A person (the subject) is tested for clairvoyance. He is shown the reverse of a randomly chosen playing card 25 times and asked which of the four suits it belongs to. The number of hits, or correct answers, is called X.

As we try to find evidence of his clairvoyance, for the time being the null hypothesis is that the person is not clairvoyant. The alternative is, of course: the person is (more or less) clairvoyant.

If the null hypothesis is valid, the only thing the test person can do is guess. For every card, the probability (relative frequency) of any single suit appearing is 1/4. If the alternative is valid, the test subject will predict the suit correctly with probability greater than 1/4. We will call the probability of guessing correctly p. The hypotheses, then, are:

- null hypothesis : $H_0 : p = \frac{1}{4}$ (just guessing)

and

- alternative hypothesis :$H_1 : p > \frac{1}{4}$ (true clairvoyant).

When the test subject correctly predicts all 25 cards, we will consider him clairvoyant, and reject the null hypothesis. Thus also with 24 or 23 hits. With only 5 or 6 hits, on the other hand, there is no cause to consider him so. But what about 12 hits, or 17 hits? What is the critical number, *c*, of hits, at which point we consider the subject to be clairvoyant? How do we determine the critical value *c*? It is obvious that with the choice *c*=25 (i.e. we only accept clairvoyance when all cards are predicted correctly) we're more critical than with *c*=10. In the first case almost no test subjects will be recognized to be clairvoyant, in the second case, a certain number will pass the test. In practice, one decides how critical one will be. That is, one decides how often one accepts an error of the first kind – a false positive, or Type I error. With c = 25 the probability of such an error is:

$$P(\text{reject } H_0 \mid H_0 \text{ is valid}) = P(X = 25 \mid p = \frac{1}{4}) = \left(\frac{1}{4}\right)^{25} \approx 10^{-15},$$

and hence, very small. The probability of a false positive is the probability of randomly guessing correctly all 25 times.

Being less critical, with *c*=10, gives:

$$P(\text{reject } H_0 \mid H_0 \text{ is valid}) = P(X \geq 10 \mid p = \tfrac{1}{4}) = \sum_{k=10}^{25} P(X = k \mid p = \tfrac{1}{4}) \approx 0.07.$$

Thus, c = 10 yields a much greater probability of false positive.

Before the test is actually performed, the maximum acceptable probability of a Type I error (α) is determined. Typically, values in the range of 1% to 5% are selected. (If the maximum acceptable error rate is zero, an infinite number of correct guesses is required.) Depending on this Type 1 error rate, the critical value c is calculated. For example, if we select an error rate of 1%, c is calculated thus:

$$P(\text{reject } H_0 \mid H_0 \text{ is valid}) = P(X \geq c \mid p = \tfrac{1}{4}) \leq 0.01.$$

From all the numbers c, with this property, we choose the smallest, in order to minimize the probability of a Type II error, a false negative. For the above example, we select: $c = 13$.

Radioactive Suitcase

As an example, consider determining whether a suitcase contains some radioactive material. Placed under a Geiger counter, it produces 10 counts per minute. The null hypothesis is that no radioactive material is in the suitcase and that all measured counts are due to ambient radioactivity typical of the surrounding air and harmless objects. We can then calculate how likely it is that we would observe 10 counts per minute if the null hypothesis were true. If the null hypothesis predicts (say) on average 9 counts per minute, then according to the Poisson distribution typical for radioactive decay there is about 41% chance of recording 10 or more counts. Thus we can say that the suitcase is compatible with the null hypothesis (this does not guarantee that there is no radioactive material, just that we don't have enough evidence to suggest there is). On the other hand, if the null hypothesis predicts 3 counts per minute (for which the Poisson distribution predicts only 0.1% chance of recording 10 or more counts) then the suitcase is not compatible with the null hypothesis, and there are likely other factors responsible to produce the measurements.

The test does not directly assert the presence of radioactive material. A *successful* test asserts that the claim of no radioactive material present is unlikely given the reading (and therefore ...). The double negative (disproving the null hypothesis) of the method is confusing, but using a counter-example to disprove is standard mathematical practice. The attraction of the method is its practicality. We know (from experience) the expected range of counts with only ambient radioactivity present, so we can say that a measurement is *unusually* large. Statistics just formalizes the intuitive by using numbers instead of adjectives. We probably do not know the characteristics of the radioactive suitcases; We just assume that they produce larger readings.

To slightly formalize intuition: Radioactivity is suspected if the Geiger-count with the suitcase is among or exceeds the greatest (5% or 1%) of the Geiger-counts made with ambient radiation alone. This makes no assumptions about the distribution of counts. Many ambient radiation observations are required to obtain good probability estimates for rare events.

The test described here is more fully the null-hypothesis statistical significance test. The null hypothesis represents what we would believe by default, before seeing any evidence. Statistical significance is a possible finding of the test, declared when the observed sample is unlikely to have occurred by chance if the null hypothesis were true. The name of the test describes its formulation and its possible outcome. One characteristic of the test is its crisp decision: to reject or not reject the null hypothesis. A calculated value is compared to a threshold, which is determined from the tolerable risk of error.

Definition of Terms

The following definitions are mainly based on the exposition in the book by Lehmann and Romano:

Statistical hypothesis

A statement about the parameters describing a population (not a sample).

Statistic

A value calculated from a sample, often to summarize the sample for comparison purposes.

Simple hypothesis

Any hypothesis which specifies the population distribution completely.

Composite hypothesis

Any hypothesis which does *not* specify the population distribution completely.

Null hypothesis (H_0)

A simple hypothesis associated with a contradiction to a theory one would like to prove.

Alternative hypothesis (H_1)

A hypothesis (often composite) associated with a theory one would like to prove.

Statistical test

A procedure whose inputs are samples and whose result is a hypothesis.

Region of acceptance

The set of values of the test statistic for which we fail to reject the null hypothesis.

Region of rejection / Critical region

The set of values of the test statistic for which the null hypothesis is rejected.

Critical value

The threshold value delimiting the regions of acceptance and rejection for the test statistic.

Power of a test ($1 - \beta$)

The test's probability of correctly rejecting the null hypothesis. The complement of the false negative rate, β. Power is termed sensitivity in biostatistics. ("This is a sensitive test. Because the result is negative, we can confidently say that the patient does not have the condition.") See sensitivity and specificity and Type I and type II errors for exhaustive definitions.

Size

For simple hypotheses, this is the test's probability of *incorrectly* rejecting the null hypothesis. The false positive rate. For composite hypotheses this is the supremum of the probability of rejecting the null hypothesis over all cases covered by the null hypothesis. The complement of the false positive rate is termed specificity in biostatistics. ("This is a specific test. Because the result is positive, we can confidently say that the patient has the condition.").

Significance level of a test (α)

It is the upper bound imposed on the size of a test. Its value is chosen by the statistician prior to looking at the data or choosing any particular test to be used. It is the maximum exposure to erroneously rejecting H_0 he/she is ready to accept. Testing H_0 at significance level α means testing H_0 with a test whose size does not exceed α. In most cases, one uses tests whose size is equal to the significance level.

p-value

The probability, assuming the null hypothesis is true, of observing a result at least as extreme as the test statistic.

Statistical significance test

A predecessor to the statistical hypothesis test. An experimental result was said to be statistically significant if a sample was sufficiently inconsistent with the (null) hypothesis. This was variously considered common sense, a pragmatic heuristic for identifying meaningful experimental results, a convention establishing a threshold of statistical evidence or a method for drawing conclusions from data. The statistical hypothesis test added mathematical rigor and philosophical consistency to the concept by making the alternative hypothesis explicit. The term is loosely used to describe the modern version which is now part of statistical hypothesis testing.

Conservative test

A test is conservative if, when constructed for a given nominal significance level, the true probability of *incorrectly* rejecting the null hypothesis is never greater than the nominal level.

Exact test

A test in which the significance level or critical value can be computed exactly, i.e., without any approximation. In some contexts this term is restricted to tests applied to categorical data and to permutation tests, in which computations are carried out by complete enumeration of all possible outcomes and their probabilities.

A statistical hypothesis test compares a test statistic (z or t for examples) to a threshold. The test statistic is based on optimality. For a fixed level of Type I error rate, use of these statistics minimizes Type II error rates (equivalent to maximizing power). The following terms describe tests in terms of such optimality:

Most powerful test

For a given *size* or *significance level*, the test with the greatest power (probability of rejection) for a given value of the parameter(s) being tested, contained in the alternative hypothesis.

Uniformly most powerful test (UMP)

A test with the greatest *power* for all values of the parameter(s) being tested, contained in the alternative hypothesis.

Common Test Statistics

One-sample tests are appropriate when a sample is being compared to the population from a hypothesis. The population characteristics are known from theory or are calculated from the population.

Two-sample tests are appropriate for comparing two samples, typically experimental and control samples from a scientifically controlled experiment.

Paired tests are appropriate for comparing two samples where it is impossible to control important variables. Rather than comparing two sets, members are paired between samples so the difference between the members becomes the sample. Typically the mean of the differences is then compared to zero. The common example scenario for when a paired difference test is appropriate is when a single set of test subjects has something applied to them and the test is intended to check for an effect.

Z-tests are appropriate for comparing means under stringent conditions regarding normality and a known standard deviation.

A *t*-test is appropriate for comparing means under relaxed conditions (less is assumed).

Tests of proportions are analogous to tests of means (the 50% proportion).

Chi-squared tests use the same calculations and the same probability distribution for different applications:

- Chi-squared tests for variance are used to determine whether a normal population has a specified variance. The null hypothesis is that it does.
- Chi-squared tests of independence are used for deciding whether two variables are associated or are independent. The variables are categorical rather than numeric. It can be used to decide whether left-handedness is correlated with

libertarian politics (or not). The null hypothesis is that the variables are independent. The numbers used in the calculation are the observed and expected frequencies of occurrence.

- Chi-squared goodness of fit tests are used to determine the adequacy of curves fit to data. The null hypothesis is that the curve fit is adequate. It is common to determine curve shapes to minimize the mean square error, so it is appropriate that the goodness-of-fit calculation sums the squared errors.

F-tests (analysis of variance, ANOVA) are commonly used when deciding whether groupings of data by category are meaningful. If the variance of test scores of the left-handed in a class is much smaller than the variance of the whole class, then it may be useful to study lefties as a group. The null hypothesis is that two variances are the same – so the proposed grouping is not meaningful.

In the table below, the symbols used are defined at the bottom of the table. Proofs exist that the test statistics are appropriate.

Name	Formula	Assumptions or notes
One-sample z-test	$z = \frac{\bar{x} - \mu_0}{(\sigma / \sqrt{n})}$	(Normal population or $n > 30$) and σ known. (z is the distance from the mean in relation to the standard deviation of the mean). For non-normal distributions it is possible to calculate a minimum proportion of a population that falls within k standard deviations for any k.
Two-sample z-test	$z = \frac{(\bar{x}_1 - \bar{x}_2) - d_0}{\sqrt{\frac{\sigma_1^2}{n_1} + \frac{\sigma_2^2}{n_2}}}$	Normal population and independent observations and σ_1 and σ_2 are known
One-sample *t*-test	$t = \frac{\bar{x} - \mu_0}{(s / \sqrt{n})}$, $df = n - 1$	(Normal population or $n > 30$) and σ unknown
Paired *t*-test	$t = \frac{\bar{d} - d_0}{(s_d / \sqrt{n})}$, $df = n - 1$	(Normal population of differences **or** $n > 30$) and σ unknown or small sample size $n < 30$

Two-sample pooled *t*-test, equal variances	$t = \dfrac{(\bar{x}_1 - \bar{x}_2) - d_0}{s_p \sqrt{\dfrac{1}{n_1} + \dfrac{1}{n_2}}}$, $s_p^2 = \dfrac{(n_1 - 1)s_1^2 + (n_2 - 1)s_2^2}{n_1 + n_2 - 2}$, $df = n_1 + n_2 - 2$	(Normal populations or $n_1 + n_2 > 40$) and independent observations and $\sigma_1 = \sigma_2$ unknown
Two-sample unpooled *t*-test, unequal variances (Welch's *t*-test)	$t = \dfrac{(\bar{x}_1 - \bar{x}_2) - d_0}{\sqrt{\dfrac{s_1^2}{n_1} + \dfrac{s_2^2}{n_2}}}$, $df = \dfrac{\left(\dfrac{s_1^2}{n_1} + \dfrac{s_2^2}{n_2}\right)^2}{\dfrac{\left(\dfrac{s_1^2}{n_1}\right)^2}{n_1 - 1} + \dfrac{\left(\dfrac{s_2^2}{n_2}\right)^2}{n_2 - 1}}$	(Normal populations or $n_1 + n_2 > 40$) and independent observations and $\sigma_1 \neq \sigma_2$ both unknown
One-proportion z-test	$z = \dfrac{\hat{p} - p_0}{\sqrt{p_0(1 - p_0)}}\sqrt{n}$	$n \cdot p_o > 10$ and $n\,(1 - p_o) > 10$ and it is a SRS (Simple Random Sample).
Two-proportion z-test, pooled for $H_0 : p_1 = p_2$	$z = \dfrac{(\hat{p}_1 - \hat{p}_2)}{\sqrt{\hat{p}(1 - \hat{p})(\dfrac{1}{n_1} + \dfrac{1}{n_2})}}$, $\hat{p} = \dfrac{x_1 + x_2}{n_1 + n_2}$	$n_1 p_1 > 5$ and $n_1(1 - p_1) > 5$ and $n_2 p_2 > 5$ and $n_2(1 - p_2) > 5$ and independent observations.
Two-proportion z-test, unpooled for $\mid d_0 \mid > 0$	$z = \dfrac{(\hat{p}_1 - \hat{p}_2) - d_0}{\sqrt{\dfrac{\hat{p}_1(1 - \hat{p}_1)}{n_1} + \dfrac{\hat{p}_2(1 - \hat{p}_2)}{n_2}}}$	$n_1 p_1 > 5$ and $n_1(1 - p_1) > 5$ and $n_2 p_2 > 5$ and $n_2(1 - p_2) > 5$ and independent observations.
Chi-squared test for variance	$\chi^2 = (n - 1)\dfrac{s^2}{\sigma_0^2}$	Normal population

Chi-squared test for goodness of fit	$\chi^2 = \sum^k \frac{(\text{observed} - \text{expected})^2}{\text{expected}}$	$df = k - 1 - \#$ *parameters estimated*, and one of these must hold. • All expected counts are at least 5. • All expected counts are > 1 and no more than 20% of expected counts are less than 5
Two-sample F test for equality of variances	$F = \frac{s_1^2}{s_2^2}$	Normal populations Arrange so $s_1^2 \geq s_2^2$ and reject H_o for $F > F(\alpha/2, n_1 - 1, n_2 - 1)$
Regression *t*-test of $H_0 : R^2 = 0.$	$t = \sqrt{\frac{R^2(n-k-1^*)}{1-R^2}}$	Reject H_o for $t > t(\alpha/2, n-k-1^*)$ *Subtract 1 for intercept; *k* terms contain independent variables.

In general, the subscript o indicates a value taken from the null hypothesis, H_o, which should be used as much as possible in constructing its test statistic. ... *Definitions of other symbols:*

• α, the probability of Type I error (rejecting a null hypothesis when it is in fact true) • = sample size • n_1 = sample 1 size • n_2 = sample 2 size • $\bar{x}$ = sample mean • μ_0 = hypothesized population mean	• s^2 = sample variance • s_1 = sample 1 standard deviation • s_2 = sample 2 standard deviation • t = t statistic • df = degrees of freedom • $\bar{d}$ = sample mean of differences	• $\hat{p} = x/n$ = sample proportion, unless specified otherwise • p_0 = hypothesized population proportion • p_1 = proportion 1 • p_2 = proportion 2 • d_p = hypothesized difference in proportion
• μ_1 = population 1 mean • μ_2 = population 2 mean • σ = population standard deviation • σ^2 = population variance • s = sample standard deviation • $\sum^k$ = sum (of *k* numbers)	• d_0 = hypothesized population mean difference • s_d = standard deviation of differences • χ^2 = Chi-squared statistic	• $\min\{n_1, n_2\}$ = minimum of n_1 and n_2 • $x_1 = n_1 p_1$ • $x_2 = n_2 p_2$ • F = F statistic

Origins and Early Controversy

Significance testing is largely the product of Karl Pearson (p-value, Pearson's chi-squared test), William Sealy Gosset (Student's t-distribution), and Ronald Fisher ("null hypothesis", analysis of variance, "significance test"), while hypothesis testing was developed by Jerzy Neyman and Egon Pearson (son of Karl). Ronald Fisher began his life in statistics as a Bayesian (Zabell 1992), but Fisher soon grew disenchanted with the subjectivity involved (namely use of the principle of indifference when determining prior probabilities), and sought to provide a more "objective" approach to inductive inference.

Fisher was an agricultural statistician who emphasized rigorous experimental design and methods to extract a result from few samples assuming Gaussian distributions. Neyman (who teamed with the younger Pearson) emphasized mathematical rigor and methods to obtain more results from many samples and a wider range of distributions. Modern hypothesis testing is an inconsistent hybrid of the Fisher vs Neyman/Pearson formulation, methods and terminology developed in the early 20th century. While hypothesis testing was popularized early in the 20th century, evidence of its use can be found much earlier. In the 1770s Laplace considered the statistics of almost half a million births. The statistics showed an excess of boys compared to girls. He concluded by calculation of a p-value that the excess was a real, but unexplained, effect.

Fisher popularized the "significance test". He required a null-hypothesis (corresponding to a population frequency distribution) and a sample. His (now familiar) calculations determined whether to reject the null-hypothesis or not. Significance testing did not utilize an alternative hypothesis so there was no concept of a Type II error.

The p-value was devised as an informal, but objective, index meant to help a researcher determine (based on other knowledge) whether to modify future experiments or strengthen one's faith in the null hypothesis. Hypothesis testing (and Type I/II errors) was devised by Neyman and Pearson as a more objective alternative to Fisher's p-value, also meant to determine researcher behaviour, but without requiring any inductive inference by the researcher.

Neyman & Pearson considered a different problem (which they called "hypothesis testing"). They initially considered two simple hypotheses (both with frequency distributions). They calculated two probabilities and typically selected the hypothesis associated with the higher probability (the hypothesis more likely to have generated the sample). Their method always selected a hypothesis. It also allowed the calculation of both types of error probabilities.

Fisher and Neyman/Pearson clashed bitterly. Neyman/Pearson considered their formulation to be an improved generalization of significance testing.(The defining paper was abstract. Mathcmaticians havc gcneralized and refined the theory for decades.) Fisher thought that it was not applicable to scientific research because often, during the course of the experiment, it is discovered that the initial assumptions about the null hypothesis are questionable due to unexpected sources of error. He believed that the use of rigid reject/accept decisions based on models formulated before data is collected was incompatible with this common scenario faced by scientists and attempts to apply this method to scientific research would lead to mass confusion.

The dispute between Fisher and Neyman–Pearson was waged on philosophical grounds, characterized by a philosopher as a dispute over the proper role of models in statistical inference.

Events intervened: Neyman accepted a position in the western hemisphere, breaking his partnership with Pearson and separating disputants (who had occupied the same building) by much of the planetary diameter. World War II provided an intermission in the debate. The dispute between Fisher and Neyman terminated (unresolved after 27 years) with Fisher's death in 1962. Neyman wrote a well-regarded eulogy. Some of Neyman's later publications reported p-values and significance levels.

The modern version of hypothesis testing is a hybrid of the two approaches that resulted from confusion by writers of statistical textbooks (as predicted by Fisher) beginning in the 1940s. (But signal detection, for example, still uses the Neyman/Pearson formulation.) Great conceptual differences and many caveats in addition to those mentioned above were ignored. Neyman and Pearson provided the stronger terminology, the more rigorous mathematics and the more consistent philosophy, but the subject taught today in introductory statistics has more similarities with Fisher's method than theirs. This history explains the inconsistent terminology (example: the null hypothesis is never accepted, but there is a region of acceptance).

Sometime around 1940, in an apparent effort to provide researchers with a "non-controversial" way to have their cake and eat it too, the authors of statistical text books began anonymously combining these two strategies by using the p-value in place of the test statistic (or data) to test against the Neyman–Pearson "significance level". Thus, researchers were encouraged to infer the strength of their data against some null hypothesis using p-values, while also thinking they are retaining the post-data collection objectivity provided by hypothesis testing. It then became customary for the null hypothesis, which was originally some realistic research hypothesis, to be used almost solely as a strawman "nil" hypothesis (one where a treatment has no effect, regardless of the context).

A Comparison Between Fisherian, Frequentist (Neyman–Pearson)

Fisher's null hypothesis testing	Neyman–Pearson decision theory
1. Set up a statistical null hypothesis. The null need not be a nil hypothesis (i.e., zero difference).	1. Set up two statistical hypotheses, H1 and H2, and decide about α, β, and sample size before the experiment, based on subjective cost-benefit considerations. These define a rejection region for each hypothesis.
2. Report the exact level of significance (e.g., p = 0.051 or p = 0.049). Do not use a conventional 5% level, and do not talk about accepting or rejecting hypotheses. If the result is "not significant", draw no conclusions and make no decisions, but suspend judgement until further data is available.	2. If the data falls into the rejection region of H1, accept H2; otherwise accept H1. Note that accepting a hypothesis does not mean that you believe in it, but only that you act as if it were true.
3. Use this procedure only if little is known about the problem at hand, and only to draw provisional conclusions in the context of an attempt to understand the experimental situation.	3. The usefulness of the procedure is limited among others to situations where you have a disjunction of hypotheses (e.g., either μ1 = 8 or μ2 = 10 is true) and where you can make meaningful cost-benefit trade-offs for choosing alpha and beta.

Early Choices of Null Hypothesis

Paul Meehl has argued that the epistemological importance of the choice of null hypothesis has gone largely unacknowledged. When the null hypothesis is predicted by theory, a more precise experiment will be a more severe test of the underlying theory. When the null hypothesis defaults to "no difference" or "no effect", a more precise experiment is a less severe test of the theory that motivated performing the experiment. An examination of the origins of the latter practice may therefore be useful:

1778: Pierre Laplace compares the birthrates of boys and girls in multiple European cities. He states: "it is natural to conclude that these possibilities are very nearly in the same ratio". Thus Laplace's null hypothesis that the birthrates of boys and girls should be equal given "conventional wisdom".

1900: Karl Pearson develops the chi squared test to determine "whether a given form of frequency curve will effectively describe the samples drawn from a given population." Thus the null hypothesis is that a population is described by some distribution predicted by theory. He uses as an example the numbers of five and sixes in the Weldon dice throw data.

1904: Karl Pearson develops the concept of "contingency" in order to determine whether outcomes are independent of a given categorical factor. Here the null hypothesis is by default that two things are unrelated (e.g. scar formation and death rates from smallpox). The null hypothesis in this case is no longer predicted by theory or conventional wisdom, but is instead the principle of indifference that lead Fisher and others to dismiss the use of "inverse probabilities".

Null Hypothesis Statistical Significance Testing

An example of Neyman–Pearson hypothesis testing can be made by a change to the radioactive suitcase example. If the "suitcase" is actually a shielded container for the transportation of radioactive material, then a test might be used to select among three hypotheses: no radioactive source present, one present, two (all) present. The test could be required for safety, with actions required in each case. The Neyman–Pearson lemma of hypothesis testing says that a good criterion for the selection of hypotheses is the ratio of their probabilities (a likelihood ratio). A simple method of solution is to select the hypothesis with the highest probability for the Geiger counts observed. The typical result matches intuition: few counts imply no source, many counts imply two sources and intermediate counts imply one source.

Neyman–Pearson theory can accommodate both prior probabilities and the costs of actions resulting from decisions. The former allows each test to consider the results of earlier tests (unlike Fisher's significance tests). The latter allows the consideration of economic issues as well as probabilities. A likelihood ratio remains a good criterion for selecting among hypotheses.

The two forms of hypothesis testing are based on different problem formulations. The original test is analogous to a true/false question; the Neyman–Pearson test is more like multiple choice. In the view of Tukey the former produces a conclusion on the basis of only strong evidence while the latter produces a decision on the basis of available evidence. While the two tests seem quite different both mathematically and philosophically, later developments lead to the opposite claim. Consider many tiny radioactive sources. The hypotheses become 0,1,2,3... grains of radioactive sand. There is little distinction between none or some radiation (Fisher) and 0 grains of radioactive sand versus all of the alternatives (Neyman–Pearson). The major Neyman–Pearson paper of 1933 also considered composite hypotheses (ones whose distribution includes an unknown parameter). An example proved the optimality of the (Student's) *t*-test, "there can be no better test for the hypothesis under consideration" (p 321). Neyman–Pearson theory was proving the optimality of Fisherian methods from its inception.

Fisher's significance testing has proven a popular flexible statistical tool in application with little mathematical growth potential. Neyman–Pearson hypothesis testing is claimed as a pillar of mathematical statistics, creating a new paradigm for the field. It also stimulated new applications in statistical process control, detection theory, decision theory and game theory. Both formulations have been successful, but the successes have been of a different character.

The dispute over formulations is unresolved. Science primarily uses Fisher's (slightly modified) formulation as taught in introductory statistics. Statisticians study Neyman–Pearson theory in graduate school. Mathematicians are proud of uniting the formulations. Philosophers consider them separately. Learned opinions deem the formulations variously competitive (Fisher vs Neyman), incompatible or complementary. The dispute has become more complex since Bayesian inference has achieved respectability.

The terminology is inconsistent. Hypothesis testing can mean any mixture of two formulations that both changed with time. Any discussion of significance testing vs hypothesis testing is doubly vulnerable to confusion.

Fisher thought that hypothesis testing was a useful strategy for performing industrial quality control, however, he strongly disagreed that hypothesis testing could be useful for scientists. Hypothesis testing provides a means of finding test statistics used in significance testing. The concept of power is useful in explaining the consequences of adjusting the significance level and is heavily used in sample size determination. The two methods remain philosophically distinct. They usually (but *not always*) produce the same mathematical answer. The preferred answer is context dependent. While the existing merger of Fisher and Neyman–Pearson theories has been heavily criticized, modifying the merger to achieve Bayesian goals has been considered.

Criticism

Criticism of statistical hypothesis testing fills volumes citing 300–400 primary references. Much of the criticism can be summarized by the following issues:

- The interpretation of a *p*-value is dependent upon stopping rule and definition of multiple comparison. The former often changes during the course of a study and the latter is unavoidably ambiguous. (i.e. "p values depend on both the (data) observed and on the other possible (data) that might have been observed but weren't").
- Confusion resulting (in part) from combining the methods of Fisher and Neyman–Pearson which are conceptually distinct.
- Emphasis on statistical significance to the exclusion of estimation and confirmation by repeated experiments.
- Rigidly requiring statistical significance as a criterion for publication, resulting in publication bias. Most of the criticism is indirect. Rather than being wrong, statistical hypothesis testing is misunderstood, overused and misused.
- When used to detect whether a difference exists between groups, a paradox arises. As improvements are made to experimental design (e.g., increased precision of measurement and sample size), the test becomes more lenient. Unless one accepts the absurd assumption that all sources of noise in the data cancel out completely, the chance of finding statistical significance in either direction approaches 100%.
- Layers of philosophical concerns. The probability of statistical significance is a function of decisions made by experimenters/analysts. If the decisions are based on convention they are termed arbitrary or mindless while those not so based may be termed subjective. To minimize type II errors, large samples are recommended. In psychology practically all null hypotheses are claimed to be false for sufficiently large samples so "...it is usually nonsensical to perform an experiment with the *sole* aim of rejecting the null hypothesis.". "Statistically significant findings are often misleading" in psychology. Statistical significance does not imply practical significance and correlation does not imply causation. Casting doubt on the null hypothesis is thus far from directly supporting the research hypothesis.
- It does not tell us what we want to know.

Critics and supporters are largely in factual agreement regarding the characteristics of null hypothesis significance testing (NHST): While it can provide critical information, it is *inadequate as the sole tool for statistical analysis. Successfully rejecting the null hypothesis may offer no support for the research hypothesis.* The continuing controversy concerns the selection of the best statistical practices for the near-term future given the (often poor) existing practices. Critics would prefer to ban NHST completely, forcing a complete departure from those practices, while supporters suggest a less absolute change.

Controversy over significance testing, and its effects on publication bias in particular, has produced several results. The American Psychological Association has strengthened its statistical reporting requirements after review, medical journal publishers have recognized the obligation to publish some results that are not statistically significant to combat publication bias and a journal (*Journal of Articles in Support of the Null Hypothesis*) has been created to publish such results exclusively. Textbooks have added some cautions and increased coverage of the tools necessary to estimate the size of the sample required to produce significant results. Major organizations have not abandoned use of significance tests although some have discussed doing so.

Alternatives

The numerous criticisms of significance testing do not lead to a single alternative. A unifying position of critics is that statistics should not lead to a conclusion or a decision but to a probability or to an estimated value with a confidence interval rather than to an accept-reject decision regarding a particular hypothesis. It is unlikely that the controversy surrounding significance testing will be resolved in the near future. Its supposed flaws and unpopularity do not eliminate the need for an objective and transparent means of reaching conclusions regarding studies that produce statistical results. Critics have not unified around an alternative. Other forms of reporting confidence or uncertainty could probably grow in popularity. One strong critic of significance testing suggested a list of reporting alternatives: effect sizes for importance, prediction intervals for confidence, replications and extensions for replicability, meta-analyses for generality. None of these suggested alternatives produces a conclusion/decision. Lehmann said that hypothesis testing theory can be presented in terms of conclusions/decisions, probabilities, or confidence intervals. "The distinction between the ... approaches is largely one of reporting and interpretation."

On one "alternative" there is no disagreement: Fisher himself said, "In relation to the test of significance, we may say that a phenomenon is experimentally demonstrable when we know how to conduct an experiment which will rarely fail to give us a statistically significant result." Cohen, an influential critic of significance testing, concurred, "... don't look for a magic alternative to NHST *[null hypothesis significance testing]* ... It doesn't exist." "... given the problems of statistical induction, we must finally rely, as have the older sciences, on replication." The "alternative" to significance testing is repeated testing. The easiest way to decrease statistical uncertainty is by obtaining more data, whether by increased sample size or by repeated tests. Nickerson claimed to have never seen the publication of a literally replicated experiment in psychology. An indirect approach to replication is meta-analysis.

Bayesian inference is one proposed alternative to significance testing. (Nickerson cited 10 sources suggesting it, including Rozeboom (1960)). For example, Bayesian parameter estimation can provide rich information about the data from which researchers can draw inferences, while using uncertain priors that exert only minimal influence on the results when enough data is available. Psychologist John K. Kruschke has suggested Bayesian estimation as an alternative for the *t*-test. Alternatively two competing models/hypothesis can be compared using Bayes factors. Bayesian methods could be criticized for requiring information that is seldom available in the cases where significance testing is most heavily used. Neither the prior probabilities nor the probability distribution of the test statistic under the alternative hypothesis are often available in the social sciences.

Advocates of a Bayesian approach sometimes claim that the goal of a researcher is most often to objectively assess the probability that a hypothesis is true based on the data they have collected. Neither Fisher's significance testing, nor Neyman–Pearson hypothesis testing can provide this information, and do not claim to. The probability a hypothesis is true can only be derived from use of Bayes' Theorem, which was unsatisfactory to both the Fisher and Neyman–Pearson camps due to the explicit use of subjectivity in the form of the prior probability. Fisher's strategy is to sidestep this with the p-value (an objective *index* based on the data alone) followed by *inductive inference*, while Neyman–Pearson devised their approach of *inductive behaviour*.

Philosophy

Hypothesis testing and philosophy intersect. Inferential statistics, which includes hypothesis testing, is applied probability. Both probability and its application are intertwined with philosophy. Philosopher David Hume wrote, "All knowledge degenerates into probability." Competing practical definitions of probability reflect philosophical differences. The most common application of hypothesis testing is in the scientific interpretation of experimental data, which is naturally studied by the philosophy of science.

Fisher and Neyman opposed the subjectivity of probability. Their views contributed to the objective definitions. The core of their historical disagreement was philosophical.

Many of the philosophical criticisms of hypothesis testing are discussed by statisticians in other contexts, particularly correlation does not imply causation and the design of experiments. Hypothesis testing is of continuing interest to philosophers.

Education

Statistics is increasingly being taught in schools with hypothesis testing being one of the elements taught. Many conclusions reported in the popular press (political opinion polls to medical studies) are based on statistics. An informed public should understand the limitations of statistical conclusions and many college fields of study require a course in statistics for the same reason. An introductory college statistics class places much emphasis on hypothesis testing – perhaps half of the course. Such fields as literature and divinity now include findings based on statistical analysis. An introductory statistics class teaches hypothesis testing as a cookbook process. Hypothesis testing is also taught at the postgraduate level. Statisticians learn how to create good statistical test procedures (like z, Student's t, F and chi-squared). Statistical hypothesis testing is considered a mature area within statistics, but a limited amount of development continues.

The cookbook method of teaching introductory statistics leaves no time for history, philosophy or controversy. Hypothesis testing has been taught as received unified method. Surveys showed that graduates of the class were filled with philosophical misconceptions (on all aspects of statistical inference) that persisted among instructors. While the problem was addressed more than a decade ago, and calls for educational reform continue, students still graduate from statistics classes holding fundamental misconceptions about hypothesis testing. Ideas for improving the teaching of hypothesis testing include encouraging students to search for statistical errors in published papers, teaching the history of statistics and emphasizing the controversy in a generally dry subject.

References

- B. S. Everitt: The Cambridge Dictionary of Statistics, Cambridge University Press, Cambridge (3rd edition, 2006). ISBN 0-521-69027-7
- Gigerenzer, G (November 2004). "Mindless statistics". The Journal of Socio-Economics. 33 (5): 587–606. doi:10.1016/j.socec.2004.09.033
- Burnham, K. P.; Anderson, D. R. (2002). Model Selection and Multimodel Inference: A Practical Information-Theoretic Approach (2nd ed.). Springer-Verlag. ISBN 0-387-95364-7
- Lehmann, E. L.; Romano, Joseph P. (2005). Testing Statistical Hypotheses (3E ed.). New York: Springer. ISBN 0-387-98864-5
- Loveland, Jennifer L. (2011). Mathematical Justification of Introductory Hypothesis Tests and Development of Reference Materials (M.Sc. (Mathematics)). Utah State University. Retrieved 01, April 2020
- Armstrong, J. Scott (2007). "Significance tests harm progress in forecasting". International Journal of Forecasting. 23 (2): 321–327. doi:10.1016/j.ijforecast.2007.03.004

- Hinkelmann, Klaus and Kempthorne, Oscar (2008). Design and Analysis of Experiments. I and II (Second ed.). Wiley. ISBN 978-0-470-38551-7
- Kruschke, J K (July 9, 2012). "Bayesian Estimation Supersedes the T Test". Journal of Experimental Psychology: General. 142: 573–603. doi:10.1037/a0029146

Permissions

All chapters in this book are published with permission under the Creative Commons Attribution Share Alike License or equivalent. Every chapter published in this book has been scrutinized by our experts. Their significance has been extensively debated. The topics covered herein carry significant information for a comprehensive understanding. They may even be implemented as practical applications or may be referred to as a beginning point for further studies.

We would like to thank the editorial team for lending their expertise to make the book truly unique. They have played a crucial role in the development of this book. Without their invaluable contributions this book wouldn't have been possible. They have made vital efforts to compile up to date information on the varied aspects of this subject to make this book a valuable addition to the collection of many professionals and students.

This book was conceptualized with the vision of imparting up-to-date and integrated information in this field. To ensure the same, a matchless editorial board was set up. Every individual on the board went through rigorous rounds of assessment to prove their worth. After which they invested a large part of their time researching and compiling the most relevant data for our readers.

The editorial board has been involved in producing this book since its inception. They have spent rigorous hours researching and exploring the diverse topics which have resulted in the successful publishing of this book. They have passed on their knowledge of decades through this book. To expedite this challenging task, the publisher supported the team at every step. A small team of assistant editors was also appointed to further simplify the editing procedure and attain best results for the readers.

Apart from the editorial board, the designing team has also invested a significant amount of their time in understanding the subject and creating the most relevant covers. They scrutinized every image to scout for the most suitable representation of the subject and create an appropriate cover for the book.

The publishing team has been an ardent support to the editorial, designing and production team. Their endless efforts to recruit the best for this project, has resulted in the accomplishment of this book. They are a veteran in the field of academics and their pool of knowledge is as vast as their experience in printing. Their expertise and guidance has proved useful at every step. Their uncompromising quality standards have made this book an exceptional effort. Their encouragement from time to time has been an inspiration for everyone.

The publisher and the editorial board hope that this book will prove to be a valuable piece of knowledge for students, practitioners and scholars across the globe.

Index

www.ingramcontent.com/pod-product-compliance
Lightning Source LLC
LaVergne TN
LVHW080910210926
848485LV00004B/626